THE EARTH'S SHIFTING AXIS

ATL Press, Inc.
Science Publishers

Frontiers in Astronomy and Earth Science, Volume 2

THE EARTH'S SHIFTING AXIS

Clues to Nature's Most Perplexing Mysteries

Mac B. Strain

ATL Press, Inc.
Science Publishers

ATL Press, Inc.
Science Publishers
P.O. Box 4563 T Station
Shrewsbury, MA 01545 U.S.A.

Library of Congress Cataloging-in-Publication Data

Strain, Mac B.
     The earth's shifting axis : clues to nature's most perplexing
mysteries/ Mac B. Strain.
               p.   cm. - - (Frontiers in astronomy and earth science : vol. 2)
     Includes bibliographic references and index.
     ISBN 1-882360-30-3 (cloth : alk. paper), - - ISBN 1-882360-31-1
(softcover : alk. paper)
     1. Earth--Rotation.   I. Title.   II. Series.
QB633.S77   1997
550' . 1 '52535--dc20   96-38267
                                                             CIP

Printed in Canada on acid-free paper ∞

PREFACE

The Earth's Shifting Axis fosters the natural curiosity about the forces that mold our planet. This curiosity is not limited to earth science professionals, although they will ultimately be the ones to pass judgment on the merits of the concepts presented here. Few activities provide a greater satisfaction and challenge to the mind than the search for insight into the marvels of science.

This book introduces the Dynamic Axis Theory. The theory addresses the role the earth spin axis played in writing the script for our planet's history and provides clues for the present and future. I invite both professionals and amateur scientists to stretch their imagination in speculating on the causal relations of Nature's master plan. Evaluate the merits of the concepts presented here based on your knowledge and logic.

Weights that shift on and within our free-spinning planet change the centrifugal forces. These, in turn, cause the axis to wobble and eventually seek a new axis of rotation. Tapping the kinetic energy of earth's rotation with each axis shift, Nature is able to accomplish the task of continually remolding our planet.

We can explain how seas invaded the continents when thick sedimentary beds were deposited, by considering how ocean waters responded to changes in the centrifugal forces of our spinning orb. A complementary lowering of sea levels occurred when submarine canyons, fjords, and guyots were sculptured. Evidence suggests that both gradual evolution and cataclysmic events were involved.

The response of magma seas to the same centrifugal forces offers answers to all types of crustal gymnastics from earthquakes and volcanoes, to running a crustal elevator and propelling tectonic plates. The Plate Tectonic Theory remains part of this equation. However, I challenge the vogue conveyor belt theory as the driving force.

Changing elevations and latitudes provide an explanation for regional climate swings responsible for continental ice sheets (ice ages) and tropical rain forests (coal ages). The notion that ice and coal ages represent global events is questioned, since a tremendous amount of heat is needed to create glaciers. Also consider that today's world climates include both extremes.

From this new perspective, you will never look at mountains, plains, oceans, deserts, and forests the same. Other puzzle pieces fall neatly into place: frozen mammoths, mass extinction events, magnetic pole reversals, hot spots, and drifting continents. Even the Colorado Plateau and Canadian Shield have much in common.

The Earth's Shifting Axis will test your patience, since it would have been much easier to introduce each concept after discussing all other interrelated concepts. Faced with this impossible task, however, it will be necessary from time to time to place some unanswered questions on hold, until the companion concepts are presented. Chapters 2 and 3 should clarify most questions.

Acknowledgements

The seeds for this avocational study were planted in the late 1930s by my Grandfather, William Chester Strain (1864-1940). He visualized the relationship between the earth's spin axis and the equatorial bulge. He used this phenomenon to develop a theory about ice ages based on axis shifts that increased latitudes and elevated lands to provide the platform on which ice sheets formed.

Late in the 1930s he had a stroke that cost him the use of his left side. During his last years, he had the patients to explain his theory to his teenage grandson who rode his bike about two and one-half miles over slightly gravel roads (not suitable for travel when wet) for a visit. The milk shake my Grandmother made (a glass of milk with some vanilla extract and ice that had been crushed in a salt sack by blows from a knife handle) may well have been the primary attraction for a visit. Many subsequent discussions with my Father, Robert Louis Strain, and Uncle David Fletcher Strain clarified points about the theory that were difficult for a teenager to comprehend.

A special thanks to two English teachers, one in high school and one in college. They challenged me by assigning a poor grade for writing assignments I had based on my grandfather's ice age theory. They faulted me for including insufficient encyclopedia or textbook references—references that are still not available. More determination to find clues to support or shoot down the ice age theory was furnished by a college professor for a course in engineering geology who simply dismissed the theory as not being in step with the vogue theories of that day—pre-plate tectonics.

I am grateful to earth scientists, too numerous to identify, in the academic, government, and private communities that I pestered with letters seeking answers to all kinds of questions. Attempts to furnish enough background on my perspective made the questions unwieldy. Without the proper perspective, their responses tended to revert to today's vogue theories. While everyone seemed reluctant to comment on areas they felt outside their area of expertise, encouragement came from their lack of serious challenges within their field specialty.

A special thanks must go to my wife for tolerating the many hours I spent on research and writing, usually with little evidence of progress.

CONTENTS

Cover design: Talieh Graphics
Cover illustration: Map of Indian Ocean seafloor gravity obtained from Geosat satellite images. Data courtesy Dr. David Sandwell, Scripps Institution for Oceanography, La Jolla, and Dr. Walter Smith, National Oceanic and Atmospheric Administration, Washington; simulated 3-D image courtesy Seabeam Instruments, Inc., East Walpole. Fledermaus software and technical support by Ocean Mapping Group, University of New Brunswick. The publisher is specially grateful to Sharon E. Spitzak of Seabeam Instruments for her assistance. The permission to reproduce this image does not constitute an endorsement of the views expressed in this book.

1   DYNAMIC AXIS THEORY

Most theories that try to explain the cause of ice and coal ages generate more questions than answers. So do the attempts to explain Nature's mechanics for raising and lowering regions, such as the Colorado Plateau and Great Plains between periods of erosion and periods, when its thick sedimentary beds were deposited. Even the vogue explanations for the horizontal forces that propel tectonic plates are open to many challenges. Is it a surprise that geology text-books fail to address the question of which continental lands and mountain ranges provided the materials for all the thick sedimentary beds of the world? Explanations for submarine canyons, mass extinctions, frozen mammoths, marine terraces, guyots, hot spots, volcanoes, and earthquakes continue to pose a challenge for earth scientists. At the same time, the role the earth's axis has played is generally overlooked. Yet, axis movements can provide us with a simple and logical explanation for these and other geologic phenomena.

The Dynamic Axis Theory, introduced in this book, is based on the changes in centrifugal forces that result from movements of the earth's axis. Without scientific evidence of polar movement, the theory would fail its most basic test. Without sufficient evidence, this book would not have been written. Geodesists who measured physical changes in our planet provided the primary exhibit. They discovered a very slight axis wobble and axis drift that confirmed some instability exists. To analyze the cause and effect of axis movements one needs to search no further than the fundamental laws of dynamics. After evaluating the ideas and arguments outlined on the following pages, you can be the judge of the earth's axis role in the evolution of our planet.

What if the earth's axis shifts?

What if the rotational axis of our planet shifted? To be more specific, what if the North Pole moved 4.32 degrees farther away from Mile High Denver, Colorado? The most obvious answer is that Denver's latitude would be about the same as the present latitudes of Tokyo, Japan, Tangier, Morocco, and Raleigh, North Carolina. Less apparent, but more impressive, are the changes in the sea level and the earth's crust resulting from the 4.32-degree axis shift.

A fictional scenario

Let us visualize the impact of the hypothetical 4.32-degree axis shift on the ocean waters of our planet. Consider the following scenario of life on the planet beginning after the short term scars of the axis shift have healed. The Denver City Council renamed their metropolis Port Denver. Ocean waters lap onto the steps of the state capital building. Ocean freighters sail up the submerged South Platte River channel and dock near the site of the old Mile High Stadium. Port Denver is the hub for cruise ships that regularly take passengers across the submerged Great Plains for a vacation in the Black Hills or Appalachian Islands. Fishing boats bring in their catch from the Louisiana Banks region of the enlarged Atlantic Ocean.

The Denver Country Club is now the Port Denver Yacht and Country Club. Palm trees line the golf course fairways, while hundreds of pleasure boats are anchored in Cherry Creek Cove. Fresh seafood restaurants, known as Fisherman's Row, line along the south shore of Golden Bay at the delta of Clear Creek. The city zoo has a large sanctuary extending to the Northeast along the inlet shore where one can view the native alligators and other subtropical plants and animals. Mass extinctions resulting from the axis shift took a toll on many plant and animal species that were previously considered part of conventional botanical garden and zoo populations.

You live in an urban development that extends to the south and east along Monument Divide between Port Denver and Colorado Springs. Recreational ocean beaches line the coast of the new North America (along the eastern base of the Rockies), from south of Carlsbad, New Mexico to north of Great Falls, Montana and into Canada. For summer enjoyment, the most popular beach in the immediate vicinity of Port Denver is found just south of Cape Limon. Cape Limon rests on the eastern tip of land extending out into the new enlarged Atlantic. During the hurricane season, Cape Limon often takes the brunt of hurricane's fury in much the same way as Cape Hatteras did before the axis shift.

On the west coast of the new North America, Pacific Ocean waters flood the lowlands. One arm of the Pacific Ocean extends up the Colorado River Basin reaching into western Colorado. The land bridge connecting North and South America no longer exists. Pacific and Atlantic waters meet in the Straight of El Paso. Colorado can no longer promote its fifty-four peaks over 14,000 feet in elevation. The highest peak, Mount Elbert, lost just less than a mile in elevation. The Rocky Mountains still offer a short ski season with heavy wet snow accumulations during the winter months as prevailing atmospheric currents provide a marked increase in annual precipitation. A warm ocean current from the western Atlantic elevates coastal temperatures along the new east coast.

The flooding of North America is matched by sea level changes in the Indian Ocean region that eliminated Australia's continental status. Only northern Australia and some of the mountain ranges remain above sea level as large islands.

Ocean waters that flood North America and Australia come from the lowered seas that were drained from coastal shelves around South America and Asia—drainage that made both continents larger and higher. Lower sea levels in South America added 3,924 feet to the 22,835 foot Cerro Aconcagua in Argentina (the highest point in South America). Mount Everest offers new challenges to avid mountain climbers with its 33,024 foot elevation. Siberia, being both higher in latitude and elevation, provides the platform for a new continental ice cap. Glacial ice extends from northern China north over the elevated continental shelf. Year around sea ice extends beyond the North Pole.

Fiction as a tool for science

The above scenario is obviously fiction dealing with imagined or invented ideas—so why is it included in a publication that tries to raise a serious point about earth science? There are several reasons. Any speculation as to exactly what will happen hundreds or thousands of years in the future, regardless of present day scientific evidence, includes an element of fiction. The word "theory" tells us we are dealing with something short of fact. Even meteorologist's long range weather forecasts contain an element of fiction. Come to think of it, some short range forecasts based on satellite monitors, a large network of ground observation stations, large high-speed computers, and the best professional judgment occasionally miss their mark.

Science fiction has always been a part of scientific progress. Few people would have taken the creators of Buck Rogers and Dick Tracy seriously, if they had attempted to present their ideas of space travel and miniaturized electronics as pure science. Both helped stimulate active minds that eventually made the dreams come true. Fiction, as a tool for science, provides the opportunity to present general concepts without having to document every bit of detail. If the same material as published a scientific theory, it will prompt challenges from every angle, thus detracting from the fundamental concepts one is trying to present.

Similar problems exist when attempting to reconstruct geologic history. Except for a few thousand years, we have no documented records from eye witnesses to tell us what really happened, when, or how it happened. Our best detective work provides only an educated guess. Even today, we have many people attempting to reinterpret and change our recorded history. Some scientists are unnecessarily cautious to introduce new ideas that challenge popular hypothesis in fear of a flood of technical challenges. Scientific studies that are excessively delayed to ensure perfection can harm the scientific community. An alternative is to openly use fiction as a science tool and admit that tremendous gaps exist in our knowledge that makes fiction an inherent part of all scientific studies. Although fiction offers greater literary license, the same extreme care in researching facts needs to be taken to insure that event scenarios truly reflect a reasonable understanding of the scientific evidence.

The hypothetical 4.32-degree axis shift used in the above scenario was selected to fit a highly recognizable region. It should never be considered as a specific prediction of the future. Similar scenarios could be developed for many other places on earth. As additional data becomes available, the Dynamic Axis Theory may someday be instrumental in developing a more probable scenario.

In the Port Denver scenario, the Great Plains are flooded by ocean waters. If there was no evidence that land a mile above sea level had ever been under ocean waters, our scenario would crumble under its weight. Geologic evidence, however, supports the notion of many large sea level changes.

Past invasions by the ocean waters

Geologists believe that ocean waters covered Colorado several times during geologic history. They say that seas invaded the Denver area about 600 million years ago. Two more major advances and withdrawals of the sea occurred some 70-270 million years ago. Sedimentary deposits made during the periods of invading seas account for most of the two-mile thick (nearly horizontal) sedimentary beds that underlie Denver and much of eastern Colorado. The only source given for the sediments is a vaguely defined Ancestral Rocky Mountains. Most sediments were deposited in shallow seas, flood plains, and lowland swampy areas. Only a relatively small percentage of the eastern Colorado Plains sediments is a product of the tertiary erosion period when the Rockies, as we find them today, were sculptured after their initial uplift. Most of the tertiary erosion found its way to the delta regions of the rivers and streams.

Fossils of the dinosaur age are abundant in some of the sedimentary bedding. Coal seams and pockets of oil offer more clues of wet and warmer past climates that produced the hydrocarbons from lush vegetation and animal decay. Climates fluctuated between periods of hot, cold, wet, and dry. They played a major role in establishing the distinctive sedimentary beds now tilted skyward along the Colorado Front Range. Detailed analysis of the sedimentary beds reveals not only major climate swings, but also variations in the deposition mechanics, changes in the source material mix, and clues to the evolution of life forms. Geologists use the abrupt changes in sedimentary deposits to define their calendar of geologic ages.

Some time before, during, or after the sea level was lowered one mile (or the plains raised one mile), intrusions of molten magma from deep in the earth's interior pushed up the Rocky Mountains. When the magma cooled, it formed igneous rocks, such as granites and basalts. It did not happen as a single geologic event, but as a series of crustal distortion periods spanning millions of years. The sedimentary beds deposited in horizontal layers, and the much older igneous basement rock on which they were deposited, were both pushed skyward. The Flatirons near Boulder, Colorado is an example of a tilted sedimentary bed. In some cases, the uplift mechanics appear to have been compressive forces like a vise that folded the crust. If, however, the primary mechanics of the Rocky Mountain formation was caused by the collision between two tectonic plates, would the mountain structure not more closely resemble an ice jam in a river or rubble pushed up by a bulldozer? In most cases, evidence from crustal distortions points to increased magma pressure from our planet's interior bowing up the crust. Some of the magma found its way up into and through the crust as volcanic eruptions. The intruding magma pushed up the crust and cooled to become part of the igneous rock core of the present mountain ranges. Earth scientists face the challenge of determining which of the igneous rocks found within the core of the Rockies (and other mountain ranges) are part of the intruding magma that pushed up the crust. Which were formed earlier and were simply elevated by new magma intrusions pushing up both the basement rocks and overlaying sedimentary beds?

Crustal rocks, in contact with the intruding magma, were converted to metamorphic rocks under heat and pressure. The overlying sedimentary beds and their igneous platforms had to be stretched to span the intruding magma blister. The crustal stretching would have opened crevices in the sedimentary beds, weak in tensile strength, to facilitate erosion. Most of the sedimentary cap that overlaid the Rockies is gone. We can only speculate on the extent of erosion that has occurred since the mountains were born. The primary evidence remaining is the tilted beds along the Colorado Front Range. Erosion has also carved canyons deep into the basement rocks to expose a cross section of the uplifted igneous and metamorphic platform for geologist to study. Each canyon conveys a slightly different story in geologic time, magma intrusion mechanics, and chemical composition.

Today's Rockies did not exist at the time the sedimentary beds of the plains were laid down. We therefore need to look elsewhere for the continental lands and mountain ranges that furnished the gravel, sand, and silt that make up their thick beds. To account for two-mile thick beds of the Great Plains to the east, the Colorado Plateau sediments to the west, and the sediments that overlaid the elevated mountain ranges, the Ancestral Rockies or other mountain ranges must have towered over the lands—but where?

Discontinuities in many of the worlds major sedimentary beds indicate that crustal elevators would have had to travel up and down several times as deposition and erosion eras alternated. Or was it the seas that came and went? Who can say that the seas will not return?

The dynamic axis challenge

This book was primarily written to show how axis movements can provide logical answers to many of the most perplexing earth science questions. It was equally important to table the Dynamic Axis Theory for the evaluation of earth scientists. The theory suggests to take a mental or written note of our present understanding of the vogue answers for each of the following ten questions. Answer the questions again when you have finished reading this book, this time based on your logic. A comparison of the two sets of answers will indicate the influence the shifting of the earth's spin axis has on our planet's evolution. My success in presenting the Dynamic Axis Theory will be measured by the differences in the two sets of answers.

1. How did oceans flood lands that are now thousands of feet above sea level?
2. How were submarine canyons and fjords, now thousands of feet below sea level, sculptured?
3. What continental lands provided the materials to form sedimentary beds that cover much of the continents and continental shelves; in several areas between five to ten miles thick?
4. Why does global cooling and warming not adequately explain ice and coal ages?
5. How can a shifting earth spin axis account for ice and coal ages, frozen mammoths, mass extinctions, and magnetic pole reversals?
6. What energy source does Nature tap to accomplish crustal gymnastics, from earthquakes and volcanoes, to propelling tectonic plates and running a crustal elevator that lowers and raises the Colorado Plateau Region between periods of sediment deposition and erosion?
7. How does the moon, sun, and axis movements team up to massage earth's crust and trigger earthquakes and volcanoes?
8. What evidence indicates earth's axis occasionally surges?
9. Why is the conveyor belt theory an inadequate explanation for propelling tectonic plates?
10. What forces are responsible for transverse fracture zones and midocean ridge offsets?

Magma seas in the context of the Dynamic Axis Theory include all materials below the brittle crust that will yield by flowing, ever so slowly, as opposed to fracturing under stress. This includes the ductile mantle as well as the more fluid outer core. Some scientists will object to the term "magma seas" because it implies a less viscous interior. "Ductal core" was considered as an alternate term, but the implication that only the earth's core is involved leaves out the critical role of the mantle. The term "mantle" also presents a restricted vision of material that can creep, but not flow like the magma erupting from Kilauea. 'Magma seas' seems more descriptive than 'mantle seas', specially if we consider the upper mantle to represent the cream that happens to be more viscous than the deeper magma.

Cyclic aspects of the Dynamic Axis Theory

The ideas of the Dynamic Axis Theory can be explained by their causal relationships. A list of hypothesis statements that constitute a substantial portion of the theory arises when one passes through the self-generating cyclic phases. Some of the following statements may seem strange from the perspective of today's vogue earth science theories. Most statements should be clearer after the second and third chapters.

· Much of the energy required to drive tectonic plates and perform crustal gymnastics derives from the kinetic energy of the earth's rotation.

· Any shift of weights on or within the rotating planet disrupts the earth's dynamic balance and causes the unrestrained axis to wobble and drift with an occasional surge.

· Any earth axis movement, or change in the earth's spin rate, alters the centrifugal forces that act on the ocean waters. The changes in centrifugal forces cause the oceans to seek a new geoid (sea level), producing both areas of flooding and drainage.

· Any movement of the planet's axis, or change in the earth's spin rate, alters the centrifugal forces that act on the magma seas. The latter affect the magma pressure exerted on the crust. They increase the pressure when the axis shift reduces the latitude (closer to the equator), and decrease it when the new latitude is greater.

· Increased magma sea pressure on the crust creates tension that opens crustal fractures. It is also responsible for the magma intrusions into the crustal cracks. These created dikes and sills, crustal blisters, known as batholith and laccoliths, mountain ranges, volcanic eruptions, and crustal spreading along midocean ridges.

· When the magma pressure is lowered below the point necessary to support the crust, the crust is shortened and subjected to compressive forces. Decreased magma pressure and crustal compression are responsible for buckling the crust into anticline and syncline overthrust belts. One crust block is pushed up over another, and subduction zones, where crust blocks are pushed back into earth's interior.

· Shear forces in the crust result from any differential in the tension and compression forces. They are responsible for plate boundary slippage, such as the San Andreas Fault along the boundary between the Pacific and North American plates, transverse faults associated with the midocean ridges, and differential movements within crustal blocks. Shear forces are responsible for all earthquakes.

· Tectonic plate movements result from crustal stretching and compression adjustments required to fit a new ellipsoidal shape of the magma seas.

· Any axis movement will result in both latitude and elevation changes. These directly affect regional climates. Flooding of some continental lands and draining of some ocean floors rearranges the land-ocean patterns. This, in turn, alters the oceanic and atmospheric currents, and causes additional regional climate changes around the world.

· Any shift of ocean waters or crustal movements resulting from axis shifts, further alters the planet's dynamic balance and causes additional axis adjustments.

· The magma seas are subjected to the same Coriolis forces that our spinning planet applies to the ocean waters. Earth's rotation places the primary circulation pattern of the magma seas, like the oceans and atmosphere, horizontal. The vertical circulation, driven by differential heat, can best be described as a secondary circulation pattern.

Modifications needed in existing working theories

Theories on any subject constitute nothing more than a passing thought, until they are evaluated by a series of logic tests. Failure to pass some of the tests will wound, but not necessarily, shoot down a theory. If an imperfect theory happens to be the best available, it may still be advanced as a working hypothesis (a hypothesis advanced as a foundation to support other theories). As new information becomes available, some working theories can be modified to reflect the latest thoughts, while other theories may fall from favor. New evidence may even put life back into discarded theories. The Dynamic Axis Theory brings several geology-related theories under review, including:

· The vogue theory that magma sea currents, acting as a conveyor belt, provide the driving force for tectonic plates.

· The notion that ocean basins are recycled as new crust is formed along the midocean ridges and subducted along deep ocean trenches. Continents are near permanent floating crustal blocks that alternately combine into supercontinents (Pangea, Gondwana, Laurasia), and subdivide into smaller continents as we find them today.

· The notion that isostatic balance establishes the MOHO depth with its thin crust under ocean basins and thicker crust under the continental land masses.

· The theory that hot spots can be used as fixed reference points for determining plate and axis movements.

· The idea that magnetic signatures, frozen into rocks as they form from molten magma, can reliably indicate the rocks' location on the planet when they were formed.

· The theory that submarine canyons, extending out from many major river deltas to deep ocean floors, were carved by underwater currents.

· The notion that ice and coal ages are products of global cooling and warming periods caused by variations in greenhouse gasses.

· The theory that ice and coal ages are caused by fluctuations in the sun's warming caused by variations in the geometry of the earth's orbit (Milankovitch Cycles)..

· The notion that evidence of ancient ice sheets found in India, Africa, and South America proves that the continents passed over or near one of the poles in their tectonic global tour.

· The idea that the volume of water stored in the North American and European ice sheet lowered sea level by an equivalent amount during the Ice Age.

· The idea that the melting of Antarctica and Greenland ice caps will raise sea levels by an amount equivalent to their water content.

· The role of increased volcanic eruptions in global cooling as a cause of ice ages.

· Louis Alvarez' theory suggesting that mass extinctions were caused by suffocation and global cooling from the dust kicked up by asteroids striking the earth.

· The theory that well preserved mammoths found in Siberia were frozen when they fell through river ice, or into glacial crevices.

Role of theories and hypotheses

Every scientific theory provides a focus for future discussion. Only after the theory has been spelled out, can anyone question and evaluate its merits point by point. Even before a theory is born, intensive questioning serves as a sounding board for its development in the exercise of trying to formulate thoughts into clear and concise words. In the process, the flags of many related hypothesis are hoisted, only to be modified or possibly shot down when weaknesses emerge.

Why-where-when-how?

Why, where, when, and how are frequently asked questions. They must be asked with the same intensity of curiosity usually attributed to a child. Children have a seemingly unlimited supply of "why" questions in response to an adult's answers. In the frustration of responding to such an onslaught of questions, we sometimes resort to untested hypotheses or declarations to terminate the discussion. This is the point in the exchange with a child where the adult resorts to the ultimate answer: "just because." Such a response is as fatal to serious scientific study, as it is in dampening the child's healthy curiosity.

Formulating the Dynamic Axis Theory

The Dynamic Axis Theory, like many other theories, was formulated in response to hundreds of questions in an attempt to provide logical answers to earth science mysteries. The initial questioning revolved around the cause of ice ages. Each question seemed to lead into other aspects of earth science. Eventually the theory developed into a comprehensive treatment of earth science that ties together causal relationships between the asthenosphere, lithosphere, hydrosphere, atmosphere, biosphere, and magnetosphere. Its primary focus shifted from answering just the ice age riddle to the role of the axis in molding the lithosphere.

The Dynamic Axis Theory addresses the force mechanics responsible for molding the lithosphere in terms of fundamental laws of physics and dynamics. It incorporates the essential elements of the plate tectonics and other vogue theories. Many of the special purpose theories or subtheories were advanced to explain individual geologic events, but failed to catch the fancy of the scientific community. They have been dusted off to gain new credibility as a subset of the Dynamic Axis Theory's larger scope.

Scientists would like to reduce the explanation of each phenomenon into a fundamental equation. Einstein succeeded by defining the volume of equations relating to the conservation of energy simply as: $E = mc^2$. If we were to reduce the Dynamic Axis Theory into a fundamental equation, it would have to be $F = ma$; where F is the centrifugal force acting on an element of our planet, m is the mass of the element being acted upon, and a is acceleration of the mass. I am afraid, however, that the $F = ma$ equation falls far short of Einstein's contribution. This formula, that has served scientists and engineers well for many years, can be found in any basic analytical mechanics text.

Testing the Dynamic Axis Theory

Testing is a never ending process. Each bit of new collected data represents a potential challenge to the basic concepts. Some modification and clarification of the theory will, undoubtedly, be needed as testing proceeds. Only if the theory can withstand in-depth critical evaluation, will it have a chance of becoming a working hypothesis.

In the case of the Dynamic Axis Theory, we are fortunate to have volumes of raw scientific data presently available to start the testing and evaluation process. Geologists have determined the dates for the formation of sedimentary deposits, ancestral climates responsible for ice and coal ages, and crustal gymnastic events from volcanic eruptions to mountain building. Computer modeling programs, utilizing existing data, can provide a good start by checking the correlation between interrelated events. Much more raw data will need to be collected to directly address the theory. Throughout our discussion I suggest new computer modeling programs and data needed for evaluating the theory.

Working theories—winners and losers

Working theories represent the building stones of scientific progress. Not all provide the shortest path to our enlightenment. A working theory is often accepted as scientific fact. When a theory is elevated to a fact, it usually has the effect of dampening scientific progress. The success of any investigation requires searching in the right place for clues, asking the right questions, and the ability to visualize causal ties between seemingly unrelated events.

Flat earth theory

The flat earth idea is an example of a theory that retarded scientific progress. The political position at the time was primarily reponsible for stymieing the freedom of scientific thought. There is evidence that several individuals felt the theory of a flat earth was flawed, but they were unable to overcome the political suppression of opposing ideas. As long as the flat earth theory enjoyed acceptance, regardless of how it was imposed, no major advancements could be made in either astronomy or the geodetic sciences. By accepting the flat earth as a working theory people were left with two options. Either the earth's expanse reached infinity (a common concept of space today), or the earth's surface was finite, and if someone ventured too far afield they would fall off the edge.

As an interesting aside, today's surveyors engage in a plane surveying technique that uses the concept of a flat earth to greatly simplify their local surveys. This is a perfectly legitimate assumption, as long as the rectangular area of the globe that encompasses the survey has at least one small dimension. State grid systems, and the military's Universal Transverse Mercator (UTM) grid, are both designed to provide an acceptable order of accuracy for a Cartesian grid reference system, as if the earth was flat. The world maps on many school classroom walls are based on the Mercator projection. A comparison of the land shapes in the polar regions on a Mercator map with a globe map will illustrate the distortions introduced when cartographers try to represent curved surfaces on a flat surface.

James Hutton

Geology struggled as an advancing science before James Hutton (1726-1797) a Scottish geologist, generally referred to as the father of geology developed the Huttonian theory. He explained the formation of rocks and the geologic history they revealed. Hutton's ideas found continued support by John Playfair Sir Charles Lyell, Sir James Hall, and others. They were key players in launching a new geology age that recognized the role of the earth's molten interior. On the opposite side of the geologic debate, a group of scientists proposed that the earth was born of water. These opposing camps planted the seeds for future arguments between the concepts of uniformitarianism and catastrophism.

Hutton's position on uniformitarianism provided a hypothesis that helped scientists recognize many ties between the past and present. Some earth scientists who have adopted Hutton's concept of gradual evolution, have also closed the door to the option that periodic cataclysmic surges of change are a normal part of evolution. As with the damper placed on the scientific progress by the flat earth theory, these scientists are forced to ignore all evidence that cannot be explained by uniformitarianism.

Hutton's concepts have been fine tuned over the years as his theory was scrutinized. Although many of Hutton's ideas are still valid today, geology has evolved to a level that would be difficult for 18th and 19th century earth scientists to recognize.

Jean Louis Rodolphe Agassiz

Jean Louis Rodolphe Agassiz (1807-1873) introduced another major advance in geological working theories in 1837. He picked up on ideas proposed earlier by James Hutton, Alexander von Humboldt, B. F. Kuhn, Jean-Pierre Perraudin, Ignatz Venetz, Jean de Carpentier, and others. Agassiz went a step further and advanced his ideas of a continental ice sheet extending from the Arctic to France. Without Agassiz' theory, it was difficult to explain the existence of huge boulders (glacial erratic) that were foreign to the lands on which they rested. It was more challenging to explain the transport system, in terms of human engineering and mechanics, than the construction of the pyramids. The Deluge flood waters were the vogue explanation.

For the next thirty years, Agassiz led the charge to gain acceptance in the scientific community for continental ice caps and their role in sculpturing the planet. When he moved to America he found additional supporting evidence of continental glaciers. The majority of experts in the scientific community did not readily accept Agassiz's concepts, although he received more support in America than in Europe. He was recognized as an expert in fish fossils, his primary field of study, but was considered an outsider in the area of general geology and glacial science. Perhaps because he openly questioned what he observed as an outsider, it allowed him to formulate his thoughts. He was not influenced by the textbook answers handed down by the subject experts. We now know that his free thinking allowed him to better visualize what the glacial scars and deposits signified.

Alfred Lothar Wegener

In the early twentieth century Alfred Wegener (1880-1930) was fascinated by the evidence of common geologic structure, coastline shapes, and fossils of plants and animal forms between the west coast of Africa and the east coast of South America. He was not, however, the first to point out the similarity of continents on both sides of the Atlantic. Edward Suess (an Austrian geologist) and Melchior Neumayr (an Austrian paleontologist) had advanced the notion of supercontinents, but with an entirely different notion on how they had formed. Frank Bursley Taylor (1860-1939), of the United States Geological Survey, presented a paper to the Geological Society of America in 1908 that suggested continents were moving apart. The experts opposed and generally ignored Taylor's concepts. Wegener was quick to adopt the same evidence to develop and support his theory of continental drift.

Wegener assembled the previously advanced ideas and played them against earth science disciplines. He visualized all the continents grouped into a supercontinent, Pangea. For this effort, Wegener is now recognized as the father of continental drift (the Wegener theory). His theory, like Agassiz', was at first met with little enthusiasm in the scientific community. Like Agassiz, Wegener was an outsider to the field of geological science. Objections to Wegener's ideas ranged from legitimate differences of opinion to public ridicule and personal attacks. A major obstacle to the acceptance of continental drift was the lack of explanations for the tremendous forces required. Although the consensus of leading scientists rejected Wegener's theory, a few supporters did not let it die out completely. His theory did not achieve the status of a working hypothesis, until about thirty years after he died on a scientific expedition deep within Greenland's ice cap.

In the early 1960s, when the plate tectonics theory gained respectability, Wegener's concept of drifting continents was incorporated as part of the theory. Unlike Agassiz' and Wegener's experience, plate tectonics was quickly advanced to the status of a working hypothesis.

Plate tectonics theory

The plate tectonics theory assumes the earth's crust is fractured into a series of plates that can move relative to their neighboring plates. A new crust is created along the midocean ridges where the plates are pulled (or pushed) apart to widen the ocean. Crust is consumed along deep ocean trenches where one plate is being subducted under another, thus reducing the ocean width. New mountains are formed where two plates collide and the crust is thrust up, or where terranes are scraped off a subducting plate and welded to the overriding plate. Plates also move relative to each other along their edges as illustrated by the San Andreas fault.

The plate tectonics theory constituted a quantum leap in our understanding of geologic sciences. As with every major new working theory, it resulted in a surge of scientific interest and general knowledge by both professional and amateur scientists. The theory is still being fine tuned as both existing data and new data are being analyzed. The Dynamic Axis Theory incorporates plate tectonics ideas and expands their scope. It rejects what has become the most popular hypothesis for the mechanics of propelling the tectonic plates. Another weakness in the tectonics theory that the Dynamic Axis Theory addresses involves recycling sediments, erosion of sedimentary deposits to form new sedimentary beds.

Dynamics Axis Theory

It is very easy to miss what we are not searching for. Evidence that supports the Dynamic Axis Theory, but does not fit into today's vogue theories, may simply have been placed on the shelf for lack of a better fate. Now is the time to dust off earlier, unexplained observations to see if they support or contradict this new theory. It is also easy to read into scientific observations what we are expecting or hoping to find. Have the diagrams of Pangea showing relatively complete continents caused us to overlook evidence of major changes within the continents? Has the concept of global cooling and warming limited the search for evidence for coexistence of ice ages (Antarctica and Greenland) and tropical forests that exist today? Even ideas generally accepted as facts need to be double checked for their validity.

Since this book represents the introduction of the Dynamic Axis Theory, it will be some time before it can be considered as a working theory. It must pass the test of time and find support within each of the scientific disciplines involved. Based on the experiences of Agassiz and Wegener, it can be anticipated that a theory presented by an outsider will not readily gain acceptance. As an engineer by education and working professional, all I can ask is to be listened to. Let unbiased testing be the judge of this theory. Only the facts and logic it support should be accepted. The theory will likely have to be modified to accommodate additional facts. However, it should not be rejected without allowing its day in court, just because it conflicts with other vogue theories.

2 DYNAMIC AXIS THEORY AND A RIGID CRUST

Sea level and the earth's rotation

No, the crust is not rigid enough to withstand pressures exerted from within our planet! It is, however, helpful to try to limit the number of variables when we analyze a problem. Treating the crust as rigid, only temporarily, illustrates the role of the earth's spin axis by assuming that only the ocean waters respond to changes in centrifugal forces. Chapter 3 will address the physical changes in the crust as it responds to the same centrifugal forces discussed below.

To address the physical changes in the earth, numbers seem to be a necessary evil. Although the dimensions used in the following paragraphs are needed to illustrate the magnitude of change, do not let them distract you from the basic concepts being presented. The magnitude of change will be reduced when we consider the centrifugal forces acting on the earth's interior.

Centrifugal forces acting on the geoid

As the earth spins on its axis at the rate of one revolution every twenty-four hours, the ocean waters are thrust toward the equator by centrifugal forces (see Fig. 2.1). The resulting geoid (shape of the earth as defined by sea level), is an oblate ellipsoid (also commonly referred to as a spheroid), with an equatorial diameter approximately 27 miles greater than its polar diameter. The polar radius (as measured from the earth's center) is 6,356,584 meters. The radius of the spheroid increases to 6,378,295 meters at the equator to form the equatorial bulge. It should be noted that the crust, without consideration for the ocean waters, also approximates an oblate ellipsoid. Appendix C presents a discussion with force diagrams of how the centrifugal forces of the earth's rotation create the equatorial bulge for the ocean waters.

What if Mount Everest was at the equator?

If the summit of 29,141 foot Mount Everest (at 6,382,295 meters from the center of the earth), was at the equator, its elevation would only be 13,434 feet above sea level (Fig. 2.2). The top of 23,300 foot Mount McKinley, Alaska, if placed at the equator, would be 35,958 feet below sea level. If the Challenger Deep in the Mariana Trench—the lowest point on the lithosphere at 35,810 feet below sea level and 6,366,322 meters from the earth's center—was at the pole, it would be 31,946 feet above sea level. Since the Challenger Deep is a depression, it would still be under water, but at the bottom of an inland lake.

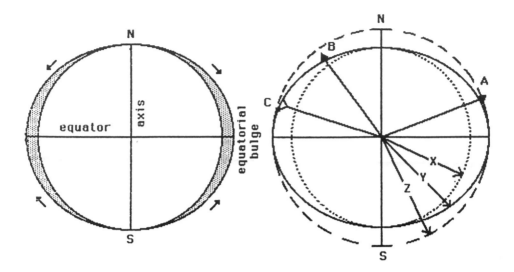

Fig. 2.1 Effect of the spinning earth's centrifugal forces on ocean waters. Centrifugal forces thrust ocean waters toward centrifugal toward the equator. The equatorial radius is about 13.5 miles greater than the polar radius.

Fig. 2.2 Effect of latitude on crustal elevations. A = Mount Everest; X = polar radius the equator; C = Challenger Deep; Z = equatorial radius; B = Mount McKinley; Y = geoid.

Squirrel cage effect

One of the fundamental laws of dynamics holds that for every action there is an equal and opposite reaction. This means that every time we take a step toward the west, we push the planet toward the east and increase its rate of spin (Fig. 2.3). Some challenging professor may some day assign a class problem to compute the squirrel cage affect assuming that everybody on earth started a marathon run toward the west at the same time (Fig. 2.4). The mathematical formulas are known. Many assumptions must be made, such as the number of people, their weights; their distribution by latitude, as well as some physical attributes of the planet. One of the major problems for the class would be to preserve enough significant figures in the calculation, so that a result can be determined. Of course, when the runners stop, the rate of spin will revert to normal—just like when the squirrel quits running in its cage.

Air currents replace the squirrel as one of the driving forces that can alter the rate of the earth's rotation. Friction between the air and crust tend to accelerate or retard the earth's spin rate. The circular motion of air currents gives the side of a high or low atmospheric cell closest to the equator a stronger influence because of its greater distance from the axis of rotation. Consider a large high pressure cell over southern Oregon. The clockwise circulation produces winds blowing west to east over the Cascade Mountain

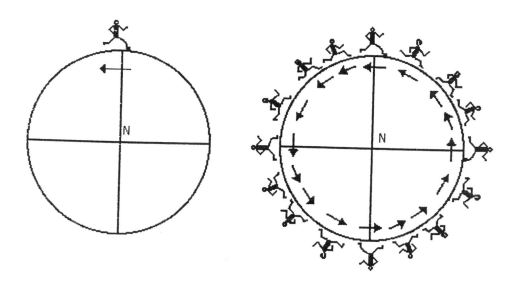

Fig. 2.3 For every force there is an equal and opposite force. Each westward step pushes the planet toward the east. Increasing, increasing its spin rate.

Fig. 2.4 Marathon runners and the squirrel cage affect. Marathon runners all running toward the west would make the earth spin faster.

Range to the North—helping the earth to spin faster. The west to east winds blowing across the Sierras to the south tend to slow the spin rate. Since the Sierras are farther from the spin axis, their slowing east to west winds have the slightly greater influence in this example. Ocean currents produce a similar effect, but were not used in this illustration since they remain relatively fixed within the confines of ocean basins.

Some of the erratic fluctuations in short term length of day observations are, undoubtedly, attributable to variations in air and ocean current patterns as they move around the globe.

Skaters spin

When weights move closer to the poles (closer to the spin axis) the earth rotates faster, as shown in Fig. 2.5. This is based on conservation of angular momentum—the same principle skaters use to spin faster by tucking their arms as close to their body as possible. When weights move toward the equator (further from the earth's spin axis), as shown in Fig. 2.6, the planet's rotation rate slows in the same way the skater terminates a spin by extending the arms and one leg. The sediments from Minnesota, deposited further from the earth's axis near the delta of the Mississippi River, will cause the earth to spin slower—increasing the length of day. The opposite is true for rivers like the Ob and Lena in the former USSR.

What if our planet stopped spinning?

To help you visualize the affect the spin rate has on the geoid, consider what would happen if the earth stopped to spin. Without the centrifugal forces acting on the seas, the gravitational forces would pull the equatorial bulge waters toward the deep polar depression to form two separated major oceans that flood the Arctic and Antarctica regions, as illustrated in Fig. 2.7. A continental landmass would ring the earth at the equator.

Small changes in the spin rate reflected in sea levels

On a more reasonable note, any decrease in the rate of axis spin will not only increase the length of a day, but also lower sea level at the equator, while raising sea level at the poles. Conversely, any increase in the spin rate would shorten the day length and increase the equatorial bulge. A change in the spin rate of one second per day would change the equatorial bulge by about twenty inches— flooding low latitude coastal region sand draining at high latitudes.

We define one day as the time it takes our planet to make a complete revolution on its axis. The day is subdivided into twenty-four hours that, in turn, are subdivided into minutes and seconds. Scientists have decided to standardized the length of a day with the aid of

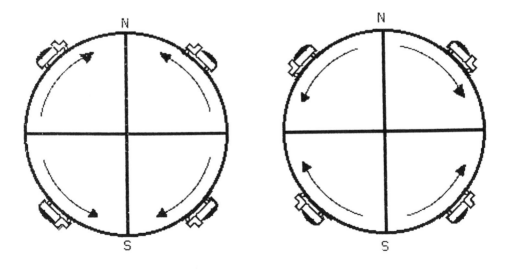

Fig. 2.5 Skaters spin concepts applied to the earth's weight shifts. Weight shifts toward the poles causes the earth to spin faster (just as a skater starts a spin then tucks the arms close to the body), decreasing the length of our day.

Fig. 2.6 Damping the skaters spin. Weights shifted toward the equator slow the rate of the earth's rotation, increasing the length of day.

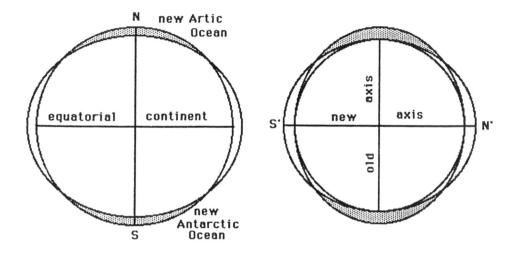

Fig. 2.7 Effect of zero spin rate. If the earth's rotation ceased, the ocean waters would form the oceans at the poles divided by a continent that rings the equator.

Fig. 2.8 Hypothetical 90 degree axis shift (assuming a rigid crust). A 90 degree shift places new pole on the equator 13.5 miles lower, and new equator 13.5 miles higher at the poles.

highly precise atomic clocks. Whenever our planet spin gets out of synchronization with the solar day, we add or subtract one second to keep our day in step with the rotation cycle, instead of redefining the length of a day (unit of time).

Since the earth's rotation rate fails to match the assigned hour, minute, second units of time, day length corrections are required periodically. Fifteen leap seconds have been added between 1972 and 1989. This does not mean that the earth has slowed by 15 seconds. Instead, the correction is similar to the day we add to February every four years to keep our year synchronized with the earth's orbiting period around the sun.

Although a decrease in the earth's spin decreases the equatorial bulge of ocean waters, the reduced weight of the waters at the equator creates a spin accelerating effect. This provides a partial built-in stabilizer on the spin rate.

Based on calculations using the fundamental laws of dynamics, the drag of the moon's gravitational pull on the tides slows the earth's rotation by one second every 62,000 years. Obviously, if the moon was the only force affecting the earth's spin rate, we could forget about making corrections to synchronize our clocks. Scientists believe that seventy-five million years ago, while the dinosaurs were still roaming the earth, the moon was about sixteen thousand miles closer to the earth. The most obvious evidence of the moon's effect on earth is the ocean tides that alter the shorelines each day. Lunar tides during the dinosaur years were larger than now. In the gravitational tug-of-war between our planet and the moon, the earth's spin rate has decreased as the moon's orbit radius increased. Any slowing of the earth's spin rate introduces a sizable variable in land-ocean distribution by altering the centrifugal force that creates the equatorial bulge of the oceans' waters. Taking the above thought a step further, about seventy-five million years ago the day was only 23 hours and 40 minutes long. At the higher spin rate, the equatorial bulge would have been about two thousand feet greater. For our rigid crust assumption, the greater equatorial bulge would have flooded equatorial lands—the extra water coming from a lower sea level in the of polar regions.

What if the poles were moved to the equator?

Although we are making some rather wild speculations about the effects changes in the earth's rotation have on the geoid, consider an axis shift of 90 degrees, as shown in Fig. 2.8. This places the new poles on the old equator, whilst the new equator passes over the old poles. Since centrifugal forces are a product of the spin axis, the old poles are flooded under ocean waters about 13.5 miles deep as the seas drain away from the new poles. The new poles are elevated by about 13.5 miles. This, of course, assumes that enough water exists to fill the new ocean basins, and that the earth's crust is rigid. We will discuss the effect of the orb's spin on the crust and asthenosphere (earth's interior) in chapter 3.

Any shift in the earth's axis or change in the spin rate will flood continents and drain ocean floors.

Geodetic reference system

In order for scientists to determine relative and absolute changes in our planet, a geodetic reference system had to be established. Several different geodetic reference systems have been designed in attempts to provide numeric addresses for each earth feature. Most measure elevations relative to sea level and horizontal positions on our planet in terms of longitude and latitude.

Geoid vertical datum

The sea level is an obvious choice for the vertical reference plane, since any tide gauge station on the coastline of continents and islands can be used to establish the data for ground surveys that extend level line elevations within the land mass. The vertical reference plane is the geoid, and is defined as the surface within and around the earth that is everywhere normal to the direction of gravity and coincides with sea level in the oceans. As an equipotential surface, it undulates over the earth's regions of high and low gravitational attractions (densities). To eliminate the variable introduced by tidal variations in sea level, the geoid is further defined as mean sea level.

The determination of mean sea level for each gauging station requires the systematic recording of tidal variations over an extended period. In the mathematical calculations of mean sea level from the tide gauge records, a proper balance between high and low tides must be used. We must take great care to eliminate secondary fluctuations of the sea due to earthquake-driven tsunamis (tidal waves), wind-driven surf, atmospheric highs and lows, or any other abnormal fluctuations of the sea level.

The Coast and Geodetic Survey (C&GS) has the responsibility of maintaining a family of tide gauge stations along all coastlines of the United States and its possessions. Similar agencies of other nations contribute to the worldwide vertical data. The C&GS is also responsible for establishing and maintaining a network of first order (very high precision) level lines with monumented bench marks which tie the tide gage stations into a geodetic network that extends the geoid datum into the continents' interior. Slightly less precise second and third order level lines further subdivide the network to provide bench marks for initiating and closing local surveys. The National Mapping Division of the United States Geologic Survey is responsible for many of the lower order surveys, as well as for producing topographic maps that show elevations in the form of contour lines.

Gradual changes in elevation have been detected in several areas. Of course, any change in elevation is relative and can be the result of a change in the geoid, vertical uplift or subsidence of the lands, or some combination of both. Resurveys that tie the area in question to stations outside the affected area are needed to determine which is the case. Land elevation changes were obvious after the 1964 Good Friday earthquake that hit Anchorage, Alaska, but it was not until extensive resurveying that the true magnitude of change was known. Subsidence near Houston, Texas is attributed to the withdrawal of oil. Uplift of Newfoundland is attributed to the crust rebounding long after the removal of the ice cap that covered Canada. Indications of lower sea levels during the last ice age are suggested to be the result of ocean waters locked in the ice cap. More discussion of coastal flooding and drainage will be presented in chapter 8.

Soon after the first satellites were launched, scientists announced that the geoid was pear-shaped and not a true spheroid. The shape was unlike a Bartlett pear, but has a girth several feet greater south of the equatorial belt line than for the corresponding latitudes of the Northern Hemisphere. Both the regional undulations and the overall pear shape of the geoid are partially attributed to the heterogeneous crust and asthenosphere (earth's interior) that create variations in the regional strengths of the gravitational attraction. Since more ocean waters exist in the Southern Hemisphere than the Northern Hemisphere also contributes to the variations in gravitational strengths by regions.

Differential leveling indirectly defines the geoid under the land masses. The C&GS conducts extensive gravimetric surveys to better define the regional and local undulations of the geoid. Satellite orbits are, likewise, sensitive to gravitational variations, providing additional worldwide data.

Spheroid horizontal datum

Although geodesists use the geoid as a reference plane for measuring elevations above and below sea level, they use a mathematically perfect, best-fit spheroid as the horizontal reference system. The geometric surface makes the spheroid pass slightly above or below the geoid to average out the undulations. Undulations of the geoid relative to the mathematical spheroid are directly proportional to gravimetric variations within the planet. We need not be concerned about the errors introduced by the selection of the reference spheroid since they need to be considered only for the most precise geodetic computations.

The spheroid is divided into degrees, minutes and seconds of longitude and latitude (Figs. 2.9 and 2.10). The origin (zero degrees) for longitude measurements is Greenwich meridian passing through Greenwich, England. The longitude values increase both east and west of Greenwich to 180-degree east and 180-degree west longitude (which are the same meridian). The origin for latitude measurements is the equator. Latitude values increase from zero at the equator to 90 degrees north at the North Pole, and 90-degree south at the South Pole.

Adopted data

With the combination of the geoid (for the vertical datum) and spheroid (the horizontal datum) as our geodetic reference system, it is possible to assign a unique positional address to each point on earth—including points above and below the earth's surface (geographic coordinates of latitude, longitude, and elevation). In 1929, the United States, Canada, and Mexico made a network adjustment of all level lines connecting the tidal stations located on the Atlantic, Gulf, and Pacific coasts to serve as the North American Vertical Datum (NAVD 29). The 1929 datum represents an adjustment of the vertical data available at one point. As additional network level lines and more precise data become available, the need for a new adjustment is indicated, but seldom accomplished. One avoids redefining the geoid, when possible, since it affects reference planes that may be used for determining relative change. A new vertical adjustment, North American Vertical Datum 1988 (NAVD 88) is now in place. Part of the differences between NAVD

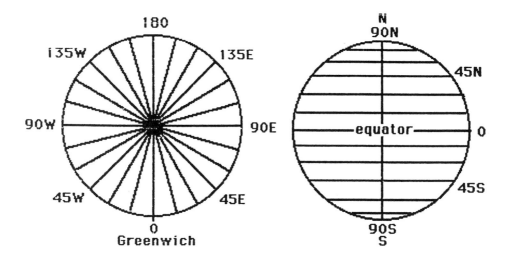

Fig. 2.9 Longitude lines (meridians) as viewed from above the North Pole Longitude lines are measured in degrees east and west of Greenwich.

Fig. 2.10 Latitude lines (parallels) as viewed from above the equator. Latitude lines are measured in degrees north and south of the equator.

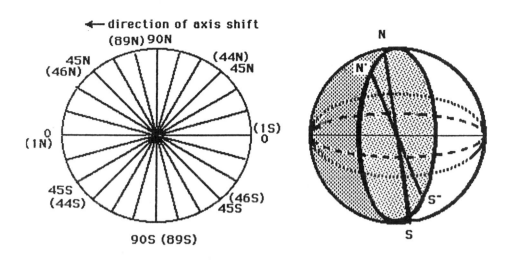

Fig. 2.11 Changes in latitude for 1-degree axis shift along the shift meridian. Latitudes increase in two quadrants, whilst decreasing in the other two quadrants. Latitude values after 1-degree axis shift shown in parenthesis.

Fig. 2.12 Dynamic axis theory special definitions. N-S'-S-N'-N = shift meridian, N-S = GRS axis, N'-S' = shift axis, heavy dash = GRS equator, fine dash = shift equator.

29 and NAVD 88 may be attributed to vertical movements of the crust. However, some differences result from the greater density of control lines and surveying accuracy. Even after new adjustments are adopted, it still takes years to convert all elevation records and update all maps to the new datum. For the conversion period, it is often necessary to either compute all old data to the new datum, or compute the new data in the old datum to determine critical relative vertical movements that span the conversion period. The above refinement is not needed to understand the Dynamic Axis Theory. It is, however, critical for interpreting the relation of changes in tide gage stations to test the theory.

Many spheroids have been invented to define the horizontal datum. Each spheroid is based on a slightly different combination of geodetic survey data. World governments have adopted specific spheroids as their official geodetic reference systems. The North American Datum of 1927, (NAD 27), based on Clarke's 1866 spheroid, was adopted as the official geodetic reference system for all of North America, Central America, Greenland, and the Philippines. A new spheroid, based on more precise and current geodetic surveys, was adopted for the 1983 North American Datum (NAD 83) update. Residual errors in the basic data available at the time Clark established his 1866 spheroid accounts for part of the inaccuracy. Part of the new best-fit-spheroid reflects physical changes in the planet (tectonic plate movements) that occurred since the geodetic data used for computing Clark's 1866 spheroid were observed. Other major spheroids include:

Clark 1880	Most of Africa, and part of the near east.
Everest	India and part of southeast Asia.
Bessel	Russia, Mongolia, Manchuria, Korea, Japan, Sumatra, Borneo, Java and Celebres.
International	Europe, China, Australia, South America and most of the ocean islands.

The U.S. Army Map Service, Corps of Engineers adopted Clark's 1866 spheroid for their worldwide Universal Transverse Mercator (UTM) Cartesian (X,Y) grid system. Elevations are given in meters above or below sea level.

A geocentric coordinate system has also been developed to define worldwide positions in Cartesian (X, Y, Z) coordinates. The geocentric reference system has many advantages if we are using a high speed computer modeling program. The (X, Y, Z) coordinates, however, adds complexity to conventional geodetic surveying—elevations measured from the earth's center makes the Panama Canal appear higher than Mount McKinley, Alaska. It is also much harder to visualize the relative positions of two points expressed in geocentric (X,Y,Z) coordinates than in longitude and latitude coordinates. Although, in the geographic coordinate system, the length of a degree of longitude decreases from its maximum at the equator to zero at the North Pole. The geocentric coordinate system does provide an ideal datum intermediary, if coordinate transformations between ellipsoids are performed.

Since the mathematical defined spheroid is established as a best-fit to the geoid and the geoid is a product of the orb's rotating axis, any changes in the earth's axis position or rate of rotation technically call for a new spheroid. Fig. 2.11 illustrates the changes in latitude for the great circle plane that contains both the geodetic reference system poles and the shift poles resulting from a 1-degree axis shift (see shift meridian discussion that follows). The values in parenthesis are the corresponding latitudes after a 1-degree axis shift.

Special or limiting definitions imposed by Dynamic Axis Theory

Several new or specially defined terms will be helpful in a discussion of the Dynamic Axis Theory (see Fig. 2.12).

The earth's axis in a dynamic axis system refers to the instantaneous axis of rotation. The earth's axis, as defined by any geodetic reference system spheroid, represents an instantaneous axis position. For our illustration we will compare latitude radii of common points on the geodetic reference system spheroid to the latitude radii of the same points on the spheroid created by our theoretical axis position.

The poles are simply the intercepts of the earth's axis with the earth's surface. As the axis moves, so do the poles. It is convenient to define axis movement with the North Pole's coordinates in the geodetic reference system. We assume a complementary South Pole position.

The equator is defined as the intercept of a plane perpendicular to the earth's axis that passes through the earth's center. Consequently, each instantaneous axis position establishes its equator.

An axis shift refers to the movement of the earth's axis between two instantaneous axis positions, regardless of the path of the poles during the axis movement. By assuming the geodetic reference system axis as one of the two axis positions, the magnitude of the axis shift is simply the difference in latitude between the geodetic reference system North Pole (90° N) and the shift pole (as defined in the geodetic reference system). Shift meridian is the great circle that passes through both the geodetic reference system pole and shift pole positions, see Figs. 2.12 and 2.13. The shift meridian represents the line of greatest latitude change resulting from an axis shift.

Axis movements

We now know that the earth's axis is in a constant state of motion known as the Chandler wobble (to be discussed in more detail in chapter 4). We also know that the crust is made up of tectonic plates in continual motion. Geodetic coordinates of stations change with plate and axis movement. Geodetic control networks that span across plate boundaries become distorted as the plates move. Unfortunately, the geodetic surveys used to establish a reference system are observed over time, which results in a slight contamination of the basic data. Redefining the spheroid, to keep up with plate and axis movements, would be a never ending task. Even if we could update it annually, we would not want to do it. By redefining the geodetic reference system, we would loose the reference base for measuring and monitoring our ever changing planet. It is fortunate that the daily and yearly changes caused by the normal plate and axis movements are so minor that they are of concern to only a handful of scientists.

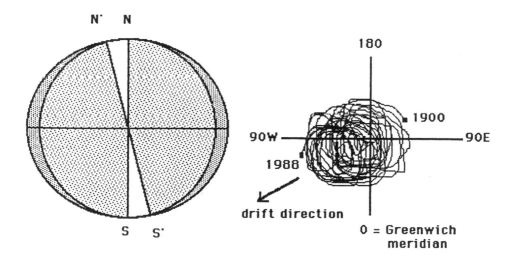

Fig. 2.13 Cross section showing shift meridian great circle. Angle SOS' = N'ON = shift angle. The shift meridian great circle passes through the GRS and shift poles.

Fig. 2.14 Stylized pole trace from 1900-1988. In 1900 the Chandler wobble pole trace over a 14-month period defined an irregular circular path around the GRS (NAD 1927) North Pole. By 1980 the 14 month trace has drifted south along about the 70°W longitude, and no longer encircles the GRS North Pole.

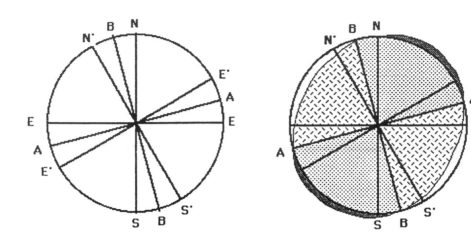

Fig. 2.16 Neutral great circles.
N-S = GRS axis; N'-S'= shift axis;
W-E = GRS equator; W'-E'=
shift equator; A-A = n e u t r a l
equatorial great circle; B-B=neutral
meridional great circle.

Fig. 2.17 Positive and negative quadrants.
Positive quadrants
    higher geoid
    lower elevations
Positive quadrants
    lower geoid
    higher elevations

For most of the following study , any one of the geoid and spheroid reference systems would be acceptable. By selectively choosing any instantaneous axis position (past or future), it is possible to calculate the change in latitude for any point on the earth. Since the radius of any point on the spheroid is a function of its latitude, the change in sea level can be determined simply by finding the difference between the latitude radius of the point on the horizontal reference system before and after the axis shift. Changes in the geoid resulting from an axis shift (assuming a rigid crust) are identical with the computed changes in the spheroid radius. To make a comparison for any point on earth between two axis positions, when neither is our horizontal reference system axis, simply determine the change in elevation points resulting from each axis position relative to the horizontal reference system, and then determine their differences.

Due to the Chandler wobble and axis drift, the North Pole traces an irregular circular path over a period of about 14 months. Fig. 2.14 shows a stylized plot of the pole trace since Chandler's discovery at the end of the nineteenth century. As evident, it resembles a very crude sample of a Palmer Method drill in penmanship ovals, initially circling the pole. Over time, the center of each circular trace gradually drifted south approximately along the 70-degree west meridian. This pole drift has been erratic in both direction and velocity.

Before we can determine the relative movements associated with geologic phenomena, we must first establish a geodetic reference system.

Theoretical one-degree axis shift

For any axis position other than that defined by the horizontal reference system, the new North Pole will be on some longitude (meridian) and at a latitude less than 90 degrees. To illustrate the affect of an axis shift on the geoid, consider the North Pole moving one degree south to the 89-degree north latitude along 90-degree east longitude (89°N, 90°E). The South Pole is assumed to have moved to its complementary position at (89°S, 90°W). For this illustration, the shift meridians are on the great circle that follows the 90-degree west and 90-degree east longitude lines. Fig. 2.15 gives the changes in the geoid radius at each 5 degrees of latitude along the shift meridian (90° west longitude) due to a 1-degree shift of the earth's axis. Table B 1 (Appendix B) lists the magnitude changes in elevation along the shift meridian (without direction), in both feet and meters, at each degree of latitude, for a 1-degree axis shift. The change in elevation is simply the difference in radius's length, as measured from the ellipsoid center, for the latitude of the point before and after the axis shift. Table B 1 gives two radii for each latitude. The radii [R] are as measured from the ellipsoid center [based on the Clark 1866 (NAD 27)], and the radii [r] are measured from the axis of rotation for the same spheroid.

Tour of shift meridian

An imaginary tour of the shift meridian for this hypothetical axis shift, will demonstrate the affect of an axis shift on the geoid using the values given in Table B 1 (Appendix B). The radius at the 90-degree latitude is 22 feet (6.6 meters) smaller than the radius at 89-degree latitude. Our hypothetical new pole at (89°N, 90°E) would, therefore, have a lower sea level (higher elevation) by 22 feet. The geodetic reference system pole would have a higher sea level (lower elevation) by 22 feet. A point midway between the horizontal reference system pole and the shift pole would be at 89.5-degree latitude, both before and after the shift. Consequently, it will have no change of elevation.

As we move south along the 90 degree west longitude shift meridian, the change in radius for a 1-degree axis shift increases to 601 feet (183.2 meters) at 75-degree north latitude, and 1,062 feet (323.8 meters) at 60-degree north latitude, both with a higher geoid (lower elevations). By the time we reach the 45-degree north latitude, the geoid is higher by 1,238 feet (377.3 meters). The point (45°N, 90°W) on this hypothetical shift meridian is just west of Wausau, Wisconsin. We could just as well have selected a different shift meridian for our illustration that would have resulted in the higher elevations being at places, such as Portland, Oregon; Ottawa, Canada; Venice, Italy; or the north tip of Japan.

Continuing south along the 90-degree west longitude, the magnitude of change per degree of shift decreases to 1,082 feet at 30-degree north latitude, and 637 feet at 15-degree north latitude. The geoid at the new equator, previously at 1-degree north latitude, would be 22 feet higher, thus making its elevation lower. At the intercept of the old equator with the shift meridian, the radius is smaller by 22 feet, giving a lower sea level and higher elevation. At the midpoint between the horizontal reference system equator and the shift equator, the latitude is 0.5 degrees both before and after the shift, thus resulting in no sea level change. It is evident, that the sea level change for the Northern Hemisphere is nearly symmetrical around the maximum at the 45-degree north latitude (minor value differences in the table are due to mathematical round-off.

Pre-shift (degrees)	Post-shift (degrees)	Change in elevation for 1-degree axis shift (feet)		
89.5N	89.5N			0
84.5N	85.5N			+172'
79.5N	80.5N		:	+424'
74.5N	75.5N		:	+620'
69.5N	70.5N		:	+796'
64.5N	65.5N		:	+949'
59.5N	60.5N		:	+1073'
54.5N	55.5N	higher elevations	:	+1161'
49.5N	50.5N		:	+1220'
44.5N	45.5N		:	+1238'
39.5N	40.5N		:	+1220'
34.5N	35.5N	lower geoid	:	+1161'
29.5N	30.5N		:	+1073'
24.5N	25.5N		:	+949'
19.5N	20.5N		:	+796'
14.5N	15.5N		:	+620'
9.5N	10.5N		:	+424'
4.5N	5.5N			+172'
0.5N	0.5S			0
5.5N	4.5S		-172'	
10.5S	9.5S		-424'	:
15.5S	14.5S		-620'	:
20.5S	19.5S		-796'	:
25.5S	24.5S		-949'	:
30.5S	29.5S		-1073'	:
35.5S	34.5S		-1161'	: lower elevations
40.5S	39.5S		-1220'	:
45.5S	44.5S		-1238'	:
50.5S	49.5S		-1220'	:
55.5S	54.5S		-1161'	: higher geoid
60.5S	59.5S		-1073'	:
65.5S	64.5S		-949'	:
70.5S	69.5S		-796'	:
75.5S	74.5S		-620'	:
80.5S	79.5S		-424'	:
85.5S	84.5S		-172'	
89.5S	89.5S			0

Fig. 2.15 Elevations (geoid) changes for 1-degree axis shift at 5-degree intervals.

The shift meridian in the Southern Hemisphere along the 90-degree west longitude is a mirror image of the Northern Hemisphere. The magnitudes of sea level change per degree of axis shift are the same, but the signs are reversed. While the Northern Hemisphere in our hypothetical axis shift experiences a flooding resulting in the lower elevations, the Southern Hemisphere experiences drainage with the lower seas. The maximum drainage occurs in the Pacific Ocean off the west coast of South America at (45°S, 90°W).

The shift meridian in the Eastern Hemisphere, along the 90-degree east longitude, will be a mirror image of the Western Hemisphere with maximum flooding at (45°S, 90°E) in the Indian Ocean and maximum drainage at (45°N, 90°E) in northeastern China. There would be no change in sea level elevation at (89.5°S, 90°E) midway between the horizontal reference system South Pole and new South Pole, or at (0.5°N, 90°E) midway between the horizontal reference system equator and shift equator intercepts.

Evidence of massive flooding and drainage of the earth's land masses during geologic history can be explained by the Dynamic Axis Theory, using small movements of the earth's axis of rotation. The changes in sea level around the world can vary from +1,238 feet to -1,238 feet for each degree of axis shift.

Neutral great circles

Besides the shift meridian great circle that contains the maximum positive and negative changes in the geoid, there are two other critical great circles we can call "neutral equatorial great circle" and "neutral meridional great circle" (Fig. 2.16). The sea level remains unchanged along these great circles since the latitudes do not change.

Neutral equatorial great circle is a great circle perpendicular to the shift meridian great circle plane that passes through the axis-of-axis pivots and bisects the horizontal reference system and shift equators (not a true equator).

Neutral meridional great circle is a great circle perpendicular to the shift meridian great circle plane that passes through the axis-of-axis pivots and midway between the horizontal reference system and shift poles (not a true meridian).

Flooded and drained quadrants

The choice of a hypothetical axis shift 90-degree east longitude for this illustration made the neutral meridian great circle close to (but not exactly on) the 0-degree and 180-degree longitude lines for small axis shifts. For small axis shifts, the neutral equatorial great circle is also close to the horizontal reference system equator. Consequently, the hemisphere divisions of the planet created by the resulting neutral great circles approximates the presently defined east, West, North and South Hemispheres.

The two neutral great circles divide the planet into four quadrants with the shift meridian down the middle of each quadrant, as shown in Fig. 2.17. Two of the quadrants have a lower sea level (higher elevations) due to the drainage resulting from a reduction in centrifugal forces. The opposite two quadrants have a higher sea level (lower elevations) due to the flooding resulting from increased centrifugal forces. In all quadrants, the maximum drainage or flooding occurs at 45 degrees of latitude along the shift meridian. The amount of drainage or flooding decreases in all directions from the quadrant maximums to zero at the neutral great circles.

For our hypothetical axis shift to (89°N, 90°E), the north half of the Western Hemisphere and the south half of the Eastern Hemisphere are flooded quadrants. The south half of the Western Hemisphere and the north half of the Eastern Hemisphere are drainage quadrants. For reasons that will become clearer in the discussion about centrifugal forces acting on the magma seas and crust (see chapter 3), we will designate the drainage quadrants as negative quadrants and the flooding quadrants as positive quadrants.

Negative quadrants are the quadrants in the direction of pole shift where the centrifugal forces are reduced, thus lowering the sea level.

Positive quadrants are the quadrants away from the direction of pole shift where the centrifugal forces are increased, thus raising the sea level.

As the axis wobble sends the North Pole in its circular path over a 14 month period, the shift meridian moves from west to east. The pattern of flooding and drainage caused by a shift in the earth's axis of rotation is divided into two positive (flooding) quadrants, and two negative (drainage) quadrants. The maximum change in elevation occurs at the center of each quadrant and decreases to zero at the equatorial and meridional neutral great circles.

Special limiting reminder

The sea level changes discussed in this chapter, will need to be modified considerably when the effect of centrifugal forces acting on the magma seas and crust are considered in chapter 3. By limiting the above discussion to a rigid crust, we have oversimplified the concepts of the Dynamic Axis Theory in an attempt to provide a better understanding of earth rotation forces.

3   DYNAMIC AXIS THEORY AND A NON-RIGID CRUST

The prediction and calculation of sea level changes resulting from any known or suspected axis shift is relatively simple and straight forward—as long as we assume that the crust is rigid. When consideration is extended to include the effect of the same centrifugal forces acting on the asthenosphere and lithosphere, we introduce a seemingly infinite number of unknowns into the equation. As discussed earlier, magma sea refers to the non-crystalline mantle and core of the earth that responds, ever so slowly, to changes in centrifugal forces. The magma seas, without crustal restraints provided by the shell, would seek an ellipsoidal equipotential surface similar to the present geoid, complete with its equatorial bulge. The resulting oblate ellipsoid would include surface undulations that result from variations in the densities of the earth's interior.

Centrifugal Forces and Magma Seas

Any shift of the earth's axis of rotation creates changes in the centrifugal forces acting on the magma interior. The magma seas, like the waters of the ocean, respond by flowing away from the negative quadrants and toward the positive quadrants. They are restricted only by their confinement within the earth's crustal shell. The thin crust—sometimes illustrated as being equivalent to the skin of an apple—is no match for the magma as it yields to the increased pressure in the positive quadrants. The crust subsides when magma pressure is reduced in the negative quadrants. Most crustal stretching and compression response occurs along seams and areas of relative weakness, such as the highly fractured junctions between existing tectonic plate edges, hot spots (local thin crustal areas), and the thinner crust under the oceans.

The most popular theory for the driving force behind the tectonic plates assumes that magma currents serve as conveyor belts to propel the plates. According to the conveyor belt theory, magma currents are driven in vertical convection loops by heat from deep within the earth's interior. Upwelling magma splits along midocean ridges to carry the plates in opposite directions. On the opposite side of the tectonic plate, where it is in collision with another plate, part of its horizontal force is either converted into a vertical force that pushes up mountains or drives the crust back into the earth's interior. Regional uplifts, e.g., the Colorado Plateau and the Great Plains—whose sedimentary beds remain nearly horizontal—do not lend themselves well to the explanation offered by basic mechanics of a magma conveyor belt driven system. The plate tectonics theory proposes that magma pressure acts on the crust. The pressure variations are a function of thermal expansion that are primarily restricted to local hot spots, upwelling along midocean ridges, and volcanic belts associated with plate subduction zones. More on the conveyor belt theory in chapter 5.

In contrast, the Dynamic Axis Theory suggests that vertical forces generated by magma sea fluctuations are the primary drivers of tectonic plates. Increased magma pressure puts the positive quadrant crust under tension as it must span the higher magma seas. Reduced magma pressure in the negative quadrants generates compressive forces sincea shortened crust can span the lower magma seas. Any differential in the tension and compressive forces that result from fluctuating magma seas, acts on the crust to create shear forces. Crustal tension, compression, and shear generated by axis movements provides a simple explanation for all the evidence of past and present crustal gymnastics. The Dynamic Axis Theory does incorporate the concept of magma currents driven by the earth's internal heat ovens, but not as the primary force for driving tectonic plates.

Classical geologic features, such as dikes, sills, batholiths, laccoliths, block mountains, igneous based mountain ranges, and volcanoes are all characteristic geologic phenomena associated with the increase of magma pressure in the positive quadrants. Synclines, anticlines, subduction zones and overthrust belts are characteristic of the geologic phenomena associated with decreased magma pressure in the negative quadrants. Both horizontal and vertical stresses responsible for crustal ruptures are caused by differential tension and compression forces acting between or within tectonic plates.

Dynamic Axis Theory and Day One of Our Planet

For the Dynamic Axis Theory to be acceptable, it must be compatible with the early days of our planet's formation. As we will see, it is relatively easy to develop a logical scenario based on the same centrifugal forces acting on today's planet Earth.

Early lithosphere

According to a frequently cited theory our planet originated as a fiery ball before cooling to form a crust with an atmosphere and hydrosphere at just the correct temperature range to support life as we know it today. Some early scientists attributed the crustal relief to the shrinking of the cooling planet. The fiery ball theory, of course, picks up on our planet's history long after the Big Bang or other theories that have been advanced for the origin of our universe. The Dynamic Axis Theory is fully consistent with a scenario that considers the formation of the earth from a fiery ball of magma.

In the Dynamic Axis Theory scenario, centrifugal forces thrust the magma seas toward the equator to create an oblate ellipsoid-shaped ball of fire. If we also assume that the axis of rotation is nearly transverse to the earth's orbit, as it is today, the coolest areas on the magma seas would be at the poles. Since the polar areas lose more heat to space by radiation than they received from the sun's radiation, the polar crust would freeze first. The heterogeneous magma, however, is made up of many compounds with widely different freezing points. Consequently, freezing of the original crust depended both on the earth's orientation to the sun and the chemical makeup of the exposed surface magma on a regional basis. Eventually a thin crust encased the entire magma ball. Boiling magma, fueled by heat from within, continued to erupt through the crust—remelting and refreezing the fragile shell. Each melt-freeze cycle added to the terrain roughness.

To set the stage for further discussion, it will be helpful to review some fundamentals of basic rock types:

Igneous rocks are formed when magma freezes. The type of igneous rock depends on several factors, including the magma's chemical composition and the cooling rate. Magma exposed to the atmosphere or hydrosphere cools rapidly to form fine grained igneous rocks. The basalts formed when the lava that pours out of Kilauea is a classic example. Magma cooled in the higher atmospheric temperatures near the equator should be slightly grainier than magma cooled faster in a cool Arctic atmosphere. Similarly, magma cooled in waters previously heated by magma intrusion should differ from that cooled in icy waters. The glass-like qualities of obsidian represent one extreme of rapid magma cooling. On the other end of the scale is the magma insulated from the atmosphere and hydrosphere by overlaying crust. The insulation allows the magma to cool slowly and form large crystals, such as found in the granite rocks. This brings up an interesting question. Would we expect the first rocks of the crust (which have never been found) to be fine grained because they were cooled in the atmosphere, or would the early atmosphere have been so thin that the magma cooled slowly to form a course grained igneous crust?

Increased magma pressure, created by axis movements and supplemented by heat generated deep within the planet, pushes magma up into and through the crust to cool as igneous rocks. The newly formed igneous rocks provide two primary aids to geologic research. Radioactive decay sets off a clock when magma freezes, and the magnetic orientation for that moment is locked into the rocks as they are formed. Geologists can establish a rock's age by comparing the ratios of radioactive rock elements to their known decay rates. The magnetic orientation provides clues as to where on earth relative to the magnetic poles the rocks were at the time of their formation.

Sedimentary rocks are the layered accumulation of sediment derived from many sources. If we assume the planet started as a molten ball that eventually cooled to form an atmosphere, hydrosphere, and lithosphere, the first sedimentary deposits on the newly formed crust would have been meteoric dust and space debris. When sufficient crust had formed to facilitate the buildup of magma pressures, periodic volcanic eruptions would have added some volcanic ash and debris to the sedimentary beds. Wind erosion would have preceded stream erosion. Stream erosion would not have been a factor until pockets of precipitated water introduced the hydrologic cycle. The earliest erosion, by either water or wind, would have been extremely slow because the exposed surface was hard igneous rocks. In time, sand served as an abrasive force to accelerate erosion. A caustic and violent atmosphere could also have played a major role in accelerating the early chemical decomposition and erosion of the surface rocks. The freeze and thaw cycle of moisture within the rock surfaces would have aided the breakup, only after the surface temperatures dropped below freezing. The earliest land-based sedimentary layers would have been thin and vulnerable to repeated erosion.

Since sedimentary beds are usually deposited in near horizontal layers they provid egeologists with an obvious signature of crustal distortion when the beds are tilted, broken, or twisted. Cambrian and later sedimentary layers provide many more clues to the earth's past evolution. Fossils, imprints, pollen, chemical composition, trapped gasses, and radioactive decay each provide part of the insight into the environment, biota, and physical changes of past ages. Limestone deposits tie their formation to prehistoric oceans. Salt beds suggest ancestral briny seas and dry lake beds. Sequential deposition and the gaps in sedimentary layers provide critical clues to the geologic past.

Today's sedimentary deposition on talus slopes, flood plains, the deltas of streams and rivers (as well as wind transported debris) comes primarily from the erosion of previously deposited sedimentary beds. This recycling of sedimentary deposits contaminates every new sedimentary bed with some geologic clues of climate and environmental conditions, not just from one or two, but dozens of erosion and deposition cycles over the life of our planet. Each newly deposited sedimentary layer includes an infusion of some freshly eroded igneous and metamorphic rocks, meteoric dust, and volcanic ash. Some of the gravel is reduced to sand, and sand is reduced to silt in each erosion cycle. Vegetation is continually altering the sediments by extracting minerals from the surface soils, as well as adding their decomposition to the mix. Groundwater, oil, and other fluids remove and re-deposit the altered chemicals in the earth's crust. Heat and pressure cause chemical change in the crustal rocks in much the same way as a baker creates pies, cakes, and breads from ingredients that are not directly recognizable in the final product.

Metamorphic rocks are formed when either igneous, sedimentary, or other metamorphic rocks are subject to high enough temperatures and pressure to partially melt and remold the rocks. Magma intrusions provide both the heat and pressure required to alter some of the rocks they come in contact with. The existence of metamorphic rocks, between the igneous rocks formed by magma intrusions and overlaying rocks, tells the story of the magma intrusions.

Early atmosphere

Some of the early atmosphere was likely pulled into the influence of the earth, as it sailed through space. Most of the atmosphere, however, was apparently derived from the gasses emitted by the molten magma. The composition of the early atmosphere would have been considerably different from today's atmosphere—probably more in tune with the gasses now being emitted from Kilauea. At high temperatures, more and more compounds are included in the vaporized mix of the atmosphere. As the earth's atmosphere cooled down, each of the vaporized compounds liquefied and solidified as they reached their critical temperatures. This would indicate that if a slice of the earliest crust could be found, it would be a lamination of compounds deposited in the order they were solidified (inverse order of their melting points).

Early Hydrosphere

Temperatures would have had to drop below the boiling point of water before the hydrosphere formed. It would have required an extended precipitation period to fill the depressions (early seas) in the newly formed crust. As noted above, the crustal relief would not have developed until the crustal thickness was sufficient to restrain the magma. Evaporation rates would have been much higher due to the warmer crust, thus speeding up the hydrologic cycle. The early crust, made up of mostly igneous and metamorphic rock, would have absorbed less of the rain waters making the streams run higher and faster. If today's volume of ocean waters precipitated out of the atmosphere before the relief in the crust formed its continents and ocean basins, the entire globe would have been covered with ocean waters. The erosion cycle would not have stepped into full gear, until larger islands and continents were created by volcanic and batholithic intrusions.

Alternate theory of the earth's origin

George W. Wetherill in his book "The Formation of the Earth from Planetesimals", suggested in an opposing theory to the fiery planet origin, that the earth formed by accumulation of cold galactic debris. According to this theory, heat within the core was created by pressure and nuclear fusion and by planetesimal collisions. Igneous, sedimentary, and metamorphic rocks that makeup the present crust would not have formed until the heat had melted the galactic materials, which then refroze into the present crust. In either case, the crust of the planet must have passed through a molten stage, making the concepts of the Dynamic Axis Theory equally compatible with a hot or cold earth origin.

Magma Pressure on the Earth's Crust

Any movement of the earth's axis will increase the magma pressure acting on the lithosphere in the two positive quadrants, while decreasing the pressure on the crust of the two negative quadrants. This is as predictable as the changes in the geoid when one assumes the crust is rigid. Positive magma pressure on the crust at any point, however, depends on the accumulated pressure changes of a series of axis movements. Crust pushed down into the magma sea, like a hydraulic plunger, adds to the pressure. So does steam produced by fluids heated by intruding magma from nuclear hot plates deep within our planet that introduce a thermal expansion of the mantle and crust. Magma intrusions into crustal cavities and volcanic discharge through the crust reduce the magma pressure. Magma viscosity also plays a part in determining how quickly magma pressure variations are stabilized between the positive and negative pressure quadrants. In other words, areas within a quadrant defined as positive or negative, based on the latest determined axis movement, can actually have zero or an inverse pressure compared to that expected.

Hydraulic principles and magma seas

When magma is confined within the lithosphere shell, any change in pressure must react according to the laws of hydraulics. If the magma was a frictionless fluid, the moment the crust was punctured, all positive pressure would be released rapidly as if a balloon or tire were punctured. However, the high viscosity of the magma greatly reduces both the reaction time and the area of direct influence. Consider the differences in responses to a puncture of water beds filled with air, water, thick syrup, or silly putty. Pressure produced by someone laying on the water bed will create completely different types of eruptions through a pin hole puncture or a slit, depending on the filling viscosity and the distance of the opening from the spot where a person is laying. It is the magma sea viscosity that prevents all magma pressure from being released through any volcanic orifice in the crust.

Because of the high magma viscosity, pressure released by the eruption of Mount St. Helens may have significantly reduced the magma pressure under Mount Adams about 35 miles to the east, while having much less effect on the magma pressure under Mount Shasta about 340 miles to the south. Of course, the size of a magma chamber that feeds the volcano will also have an influence on regional pressure distribution.

The Dynamic Axis Theory offers a logical explanation for variations in vertical pressure exerted by the magma seas on the crust. All the past and present volcanic activity, as well as crustal uplifts provide extensive evidence that the pressure of the magma seas is not constant. It is also much easier to account for magma intrusions directly from magma pressure variations created by axis movements, than from a secondary reaction responding to horizontally driven tectonic plates.

Variable Great Circle Lengths

One way of viewing the forces that act on the crust due to an axis shift, is to consider the changes in spheroid's great circle half-lengths resulting from geoid changes. Consider what happens for a family of great circle lengths that intercept the shift meridian at different latitudes.

Great circles are often thought as the shortest air route between two points—the surface intercept of a plane passing through the two points and the earth's center. Whenever the two points are 180 degrees apart, an infinite number of great circles is defined. If the two points happen to be the North and South Poles, these great circles are called meridians. When the two points are on the equator and 90 degrees from the shift meridian, an infinite number of great circles is defined with each intercepting the shift meridian at a different latitude. Each great circle across a quadrant has a different length. For all point pairs that do not fall on the great circles noted above, only one great circle is defined.

Fig. 3.1 shows the change in great circle half-lengths for a 1-degree axis shift at each 15 degree intercept along the shift meridian. The positive values indicate the magnitude of

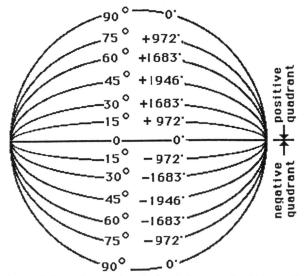

Fig. 3.1   Changes in semi-circumference length for 1-degree axis shifts at 15-degree interval meridians.

stretching with the crust under tension. The negative values indicate the amount of shortening with the crust under compression. Table B-2 (Appendix B) provides both the great circle half-lengths for each 1-degree intercept, and the changes in hemispheres great circle half-lengths resulting from a 1-degree shift of the earth axis at each degree of latitude along the shift meridian. In both cases the values are expressed as half the total great circle circumference to represent the change across the positive or negative quadrants.

To illustrate the effect of variable forces acting on the crust, the diagrams show a block of lithosphere with an upper crust of sedimentary beds and lower crust of igneous and metamorphic rocks floating on a magma sea. Fig. 3.2 depicts a block segment of the lithosphere in equilibrium with no external forces acting on it. Absolute equilibrium is rare within tectonic plates, since all plates are being jockeyed over the magma seas. Because the sedimentary bed thickness varies from zero to over 55,000 feet, and the igneous basement varies from zero (where magma is being extruded) to over 150,000 feet thickness, it would be unrealistic to try to depict all possible combinations. Also, the formation of each geologic feature in Nature is dependent on so many different variables that—like snow flakes—no two exact copies have been found. Although some reference is made to specific geologic formations, the primary objective is to present a more generic concept of geological events as they relate to the Dynamic Axis Theory. All diagrams in this text, like most geologic illustrations, constitute an oversimplification. They are usually designed to address specific points, whilst ignoring other known variables.

Crustal Tension Forces

Fig. 3.3 illustrates the typical forces acting on a crustal block segment in the positive quadrants. Increased vertical pressure, supplied by the higher magma seas, elevates the crust

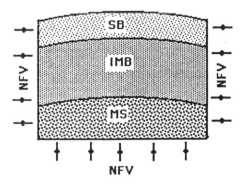

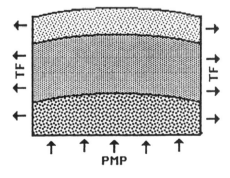

Fig. 3.2 Lithosphere block segment in equilibrium.
SB = sedimentary bedding
IMB = igneous and metamorphic basement
MS = magma seas
dot = neutral force vector

Fig. 3.3 Lithosphere block segment typical of positive quadrants.
TF = tension forces on crust;
PMP = positive magma pressure rock; arrow = force vector (direction only).

and creates horizontal tension forces in it. Changes in the family of great circle half-lengths across the positive quadrant, identified by their intercept with the shift meridian, increase from zero at the equator to a maximum at 45 degrees latitude; then decrease from 45 degrees latitude on to the pole. From Fig. 3.1, we know that the maximum change in great circle half-length is 1,946 feet per degree of axis shift. This means that the crust in the positive quadrants is under tension since the crust must stretch by 1,946 feet to span the increased great circle lengths. A non-elastic crust accommodates expansion by fracturing, thus providing the openings for magma intrusions. Volcanic eruptions will reduce the magma pressures as does magma that pushes up into the crust to form of dikes, batholiths, etc. The total volume of the higher magma seas that elevate the crust will approximate the volume of the higher geoid in the positive quadrant. This includes the magma that works its way up into or through the crust.

Crustal tension within a tectonic plate is soon dissipated since the crust has extremely poor tensile strength and poor elasticity. If we could hitch a giant locomotive up to the North American Plate and start pulling, it would resemble a railroad locomotive trying to pull a string of box cars with only their air hose connections fastened. The plate would simply separate along the first fracture zone behind the locomotive. Minute crustal thinning will occur under tensile forces. The crust is also subjected to local tension forces over areas of intruding magma that cause thermal expansion in the mantle and crust. Crustal

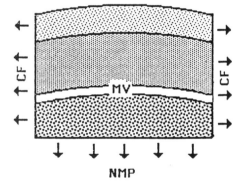

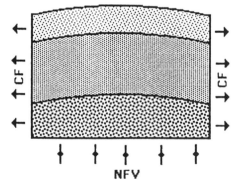

Fig. 3.4 Lithosphere block segment typical of negative quadrants.
MV = Magma void created when magma seas flow away from the negative quadrants.
NMP = Negative magma pressure created by the lowered magma seas (vacuum).

Fig. 3.5 Same as Fig. 3.4, but after the crust has subsided to rest on magma seas.
Note: When opposing compression forces are unequal, the crust must overcome the frictional resistance between the crust and magma seas before the crust will move horizontally (tectonic plate movements).
CF = Compression forces exerted on the lithosphere as it must fit into a shorter crustal segment.
MP = Magma pressure at zero after crustal subsidence.

Top: Mato Tepee and the Crown of Cassiopeia, Devils Tower National Monument, Wyoming. Courtesy Gary W. Szelc. Bottom: Devils Postpile, National Monument, California. Courtesy Mike Schulist. Overleaf: Map of Earth's seafloor gravity obtained from Geosat satellite images. Areas of high gravity are in yellow and orange, low in blue and purple. Courtesy Dr. David Sandwell, Scripps Institution for Oceanography, and Dr. Walter Smith, National Oceanic and Atmospheric Administration.

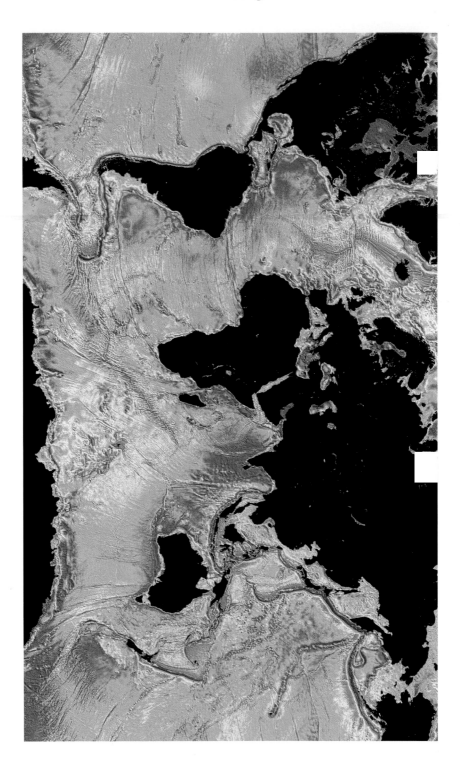

gymnastics created by increased magma pressure and crustal tension is discussed later under "Classic geologic features of the positive quadrants."

Crustal compression forces

Figs. 3.4 and 3.5 illustrate the typical forces acting on crustal blocks in the negative quadrants. The lowered magma seas create a void (negative pressure could be thought of as a temporary vacuum) that the crustal block attempts to bridge. The crust, however, does not have the strength to bridge large area voids. The crust will subside to the point where it rests on the lowered magma sea as shown in Fig. 3.5. If the block of crust subsides to rest completely on the magma seas, the so-called negative magma pressure is eliminated. The crustal compression, however, will still act on the block as it is forced into a shorter great circle.

Table B-1 (Appendix B) lists the theoretical vertical changes in the geoid and demonstrates that a 1-degree axis shift creates a maximum void depth at 45 degrees latitude of about 1,238 feet. Before the crust at 45 degrees latitude on the shift meridian can rest on the lowered magma seas, its great circle half-length must be reduced by 1,946 feet. The crust, attempting to bridge the void, is under compression similar to an arched roof. The span length and weight of crustal bridging, not the height of the void, trigger crustal failure. Under compression, such failures create some type of buckling or shear fracturing as discussed later under "Classic geologic features of the negative quadrants."

Compression forces are readily transferred within the crust. In the locomotive analogy, it would make no difference, if the boxcar connections were set—the locomotive would push all the cars. In the earth's crust, if the Atlantic floor pushes against South Carolina, the compression force (fewer friction losses) is, in turn, transferred to Georgia, Alabama, Mississippi, Louisiana, Texas, New Mexico, Arizona, and California, where it comes into play on the San Andreas fault.

Some thickening of the crust will result from crustal compression. Thermal contraction of the mantle and crust, as intruding magma cools, will result in additional crustal subsidence and compression.

Crustal shear forces

Whenever changes in magnitude of the tension or compression forces occur across a crustal block, the differential force creates shear forces that (if large enough) can rupture the crustal block. Three major types of shear forces are discussed in the following paragraphs. Fig. 3.6a represents the shear forces between two tectonic plates side by side where one block is moving horizontally relative to the other (such as the San Andreas Fault).. Fig. 3.6b illustrates the relative vertical shear movement between crustal blocks subject to differential vertical pressures (block mountain uplifts).. Fig. 3.6c features the horizontal shearing between upper and lower blocks being driven in opposite directions (overthrust belts).. Of course, crustal fractures will seldom fit any of these three examples, but they will have components of each..

The change in length of the great circle that intercepts the shift meridian at 45 degrees latitude, as mentioned above, is 1,946 feet for each degree of axis shift. At 40 degrees latitude, the change is 1,919 feet. This differential of 27 feet in great circle length induces

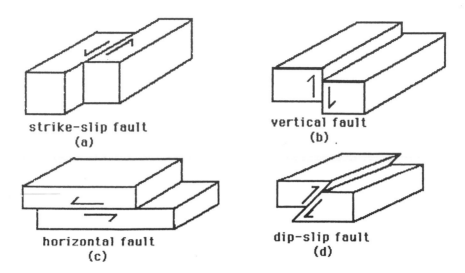

Fig. 3.6   Shear forces between crustal blocks: (a) strike-slip fault (b) vertical fault (c) horizontal fault (d) slip-dip fault..

a shearing force parallel to the great circles and between the two great circles. Direct evidence of such shearing forces acting on the crust can be seen in the transverse faults perpendicular to the midocean ridge.

Initial crustal fractures require the buildup of a greater stress than is required for a slippage involving an existing fracture. Consequently, shear forces, derived either from differential tension or compression forces usually slip along existing crustal fractures. Before a shearing force can create a slippage over an existing fault, the stress must overcome the friction lock that holds the blocks together. Appendix D provides a more detailed discussion of force vectors involved in shear movements.

Classic Positive Quadrant Features

Many geologic features can be explained by substantially increased magma pressure acting on the crust from below. When the lithosphere is stretched over the higher magma seas of the positive quadrants, the crust's poor crustal strength facilitates magma intrusions.

Dikes and sills

Even a slight increase in the magma pressure will result in molten magma pushing its way up into vertical and horizontal fractures in the crust, especially since crustal tension tends to open the crustal fractures. Under increased pressure, magma intrusions serve as wedges to expand and create additional crustal fractures. Geologists refer to the intrusions of magma into narrow crustal fractures as dikes (vertical or near vertical fractures), and sills

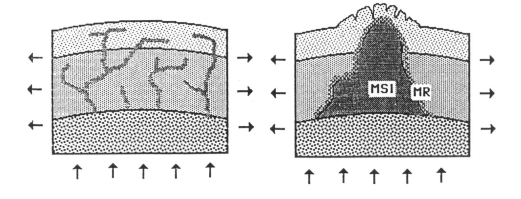

Fig. 3.7 Dikes and sills. The lithosphere under tension opens crustal fractures that are filled by the higher magma seas. The pressure of magma also spreads and extends the crustal fractures.

dikes = near vertical intrusions
sills = near horizontal intrusions

Fig. 3.8 Batholith intrusion forming crustal blister with surface crevices.

MSI = Magma stock intrusion that torches its way up into the lithosphere to create a crustal blister.

MR = Metamorphic rocks formed when heat and pressure of the intruding magma partially melt the adjacent rocks that are molded into new rocks.

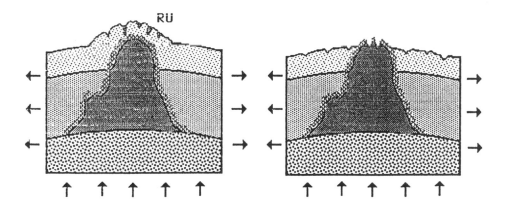
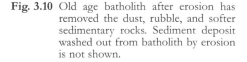

Fig. 3.9 Batholith intrusion with dust and rubble filling the crustal crevices.

RU = Rubble from the brittle sedimentary rocks and unconsolidated soils that crumble and fall into the crustal crevices opened by the batholith uplift.

Fig. 3.10 Old age batholith after erosion has removed the dust, rubble, and softer sedimentary rocks. Sediment deposit washed out from batholith by erosion is not shown.

(horizontal or near horizontal fractures). Fig. 3.7 illustrates the magma moving up into the fractures in the igneous basement rocks, and occasionally extending up into the overlaying sedimentary beds. The magma that pushes up into fractures is still well insulated from the atmosphere and hydrosphere by the overlaying crust. Consequently, the magma cools slowly to form the coarse crystalline igneous type rocks. Cross sections of dikes and sills, exposed by years of erosion, are Nature's art work of multicolored streaks that decorate the walls of many canyons and valleys.

Dikes and sills provide additional hints about the magma interior. The crossing of dikes and sills shows that magma intrusions occurred at different time intervals with each event creating a new series of crustal fractures. The variation in color, texture, and chemical composition reminds us that the magma is not homogeneous. The variations can easily be explained, if we think of the magma sea as a giant marble cake mix being continually stirred. For each magma intrusion, magma from a different part of the mix comes up through the fractures. Variations in all basement rock can be accounted for with the same reasoning. Heat and pressure from intruding magma melt some of the existing crust, and consitute another variable that affects the local mix. The melted crust becomes part of the intruding magma and alters its chemical composition. Intrusions of hot magma serve to weld the fractures, consequently, subsequent magma intrusions must find new fractures.

Batholiths and laccoliths

Magma intrusions into the crust may also take the form of near circular or elliptical blisters. The large magma domes that push up the surface rocks form what geologists call batholiths and laccoliths. Fig. 3.8 illustrates how a batholith can be formed when the magma intrusion originates as a large stock directly from the earth's interior.

The surface rocks are stretched over the intruding magma to open dike fractures in the lower surface and deep crevices in the stretched upper surface. Since any overlaying sedimentary surface beds will vary from brittle sandstone to soft shale (both weak in tensile strength), much of the fractured sedimentary beds simply collapse into the crevices as dust and rubble (Fig. 3.9). Erosion is accelerated because the batholith's highly fractured sedimentary cover is also elevated (Fig. 3.10).

The diagrams in Figs. 3.8-3.10, show the magma intrusion penetrating the sedimentary bedding. In many, if not most, cases the intrusion will terminate within or below the igneous basement rock. When this happens, the igneous basement rocks are simply bowed upward along with the sedimentary beds. A reduction in the pressure of intruding magma will occur, if the axis moves in a direction that increases the regional latitude; a nearby volcanic eruption releases pressure; or pressure is reduced as the magma blister expands.

The intense heat and pressure along the interface between the magma intrusion and the existing crust cause some of the adjacent rocks to partially melt and be remolded into metamorphic rocks. The test for penetration of the magma intrusion into the sedimentary bedding is simply the existence of metamorphism. If the sedimentary rocks overlaying an apparent igneous intrusion are not metamorphosed, the igneous rocks were cold at the time of sediment deposition. The sedimentary layers now exposed in the Grand Canyon and Glenwood Canyon on the Colorado River were both laid down on a cold pre-existing basement rock platform. When we enter Glenwood Canyon from the east, only sedimentary rocks are visible. The igneous basement rocks make their appearance at the river and road level just a few miles into the canyon. By the time we reach the Hanging Lake turnoff, about mid-canyon, the igneous/sedimentary boundary is well up on the wall of the canyon. As we continue west through the canyon, we see that the igneous/sedimentary interface again drops below the river level. Near Glenwood Springs, the crustal gymnastics becomes far more complex with sharply tilted and twisted sedimentary beds.

High pressure magma responsible for a batholith intrusion will also work its way into dike and sill fractures. If no dikes, sills, or metamorphism are evident within the overlaid sedimentary layers, the magma bubble responsible for the batholith uplift remains deeper within the crust. The Idaho Batholith, of central Idaho and western Montana, is believed to have formed (frozen) while about 15 miles below the surface. By some combination of crustal gymnastics and erosion, much of the overlaying rock has since been removed to expose the batholith structure. The Boulder Batholith, between Butte and Helena, Montana, was likely formed much closer to the surface. Volcanic activity, associated with the formation of the Boulder Batholith, created the Elkhorn Mountains as a part of the batholith intrusion.

Laccoliths are formed when magma works its way up through an orifice vent before spreading out along a horizontal seam to form the crustal blister. Fig. 3.11 illustrates the formation of a laccolith when the magma spreads out between the basement rock and overlaying sedimentary bedding. Fig. 3.12 shows the magma intrusion spreading out between sedimentary beds. From the upper surface, before erosion has exposed the complete structure, batholith and laccolith signatures are identical. The LaSal and Henry Mountains in eastern Utah are just two examples of laccolith intrusions.

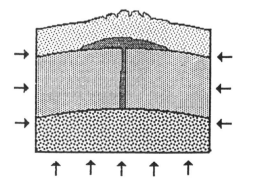
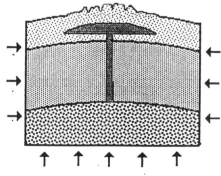

Fig. 3.11 Laccolith intrusion with magma blister formed between the basement rocks and the sedimentary bedding.

Fig. 3.12 Laccolith intrusion with magma blister formed between sedimentary bedding.

Mountain ranges

The formation of an igneous base mountain range can be explained by a scenario that combines the above explanations for dikes and batholiths. The tilted sedimentary beds that form the Flatirons near Boulder, the Red Rocks near Morrison, and the Garden of the Gods near Manitou Springs along the Colorado Front Range in Colorado, are all classic examples of the warping caused by magma intrusions along a linear rupture of the crust. The Rocky Mountains were formed by a series of magma intrusion, involving all the classic types, and extending over millions of years. We can think of the intrusion event responsible for an individual mountain range as simply oversized dikes that buckle the crust upward. In the case of the Rocky Mountains, most ranges run north-south. Fig. 3.8 could be re-labeled as a cross section of a mountain range formed by a magma intrusion along a linear rupture.

It is the variation in strengths of upturned sedimentary beds that leaves hogbacks standing parallel to the mountain range after the softer layers have been eroded away. The same sedimentary beds exposed along the eastern side of the Rockies can be found tilted up on the western side of the continental divide. If we draw a cross section and project the two mile thick sedimentary beds across the Rockies, we can visualize the mass of crust that has apparently been eroded away to expose the Rocky Mountains as we see them today. The Rockies are considered to be relatively young in geologic time, yet most of the indicated two-mile thick sedimentary cap over the Continental Divide is missing. We can only speculate how much more of the original uplift has been eroded away. The major east-flowing canyons along the Colorado Front Range, such as Arkansas, South Platte, Saint Vrain, Big Thompson, and Cache la Poudre Rivers, as well as the many creeks, have not only removed the cap, but cut deeply into the igneous and metamorphic basement rocks. Was a two-mile thick bed of sediment pushed up, and then eroded away? The major clue we have is the other thick sedimentary beds in North America that had to receive their materials from some source. If not, in part, from a sedimentary blanket overlying the Rockies, then what was their source?

If you view the Rocky Mountains and Continental Divide from the plains of eastern Colorado, visualize a horizontal line in the sky at twice the height of the tallest mountain. The imaginary line would represent the approximate height of a two-mile thick sedimentary bed cap over the Continental Divide. That would have left the Rockies a few thousand feet lower than the Himalayas compared to today's geoid. A lower geoid would have forced the Rockies even higher.

Further west of the Continental Divide, we can find that many related thick sedimentary beds have survived the erosion process, including: the Vail Pass area along Interstate I-70, Maroon Bells near Aspen, and the eastern side of Rabbit Ears Pass on US Highway 40.

One of the major shortcomings of using Figs. 3.8-3.10 as cross section diagrams is that they lack the third dimension. The figures indicate that the valleys associated with a mountain range uplift are parallel to the ridge. Natural drainage of the range, however, dictates that valleys be carved at right angles to the ridge. Actually, in the Rockies we can find many examples where segments of mountain valleys are parallel to the primary ridge. These north-south valleys now feed into the east-west flowing streams that carved the mountain drainage system. Before the river and stream system was fully developed, the parallel crevices created by the uplift may well have formed many long north-south mountain lakes. Horsetooth Reservoir, west of Ft. Collins, Colorado is man's reconstruction of how such lakes may have appeared.

The width and shape of the intrusions, as well as the spacing between intrusion faults, set the stage for the formation of Colorado's high mountain parks (wide valleys) that are often parallel to the main ridge alignment. The complex system of ranges and batholiths that formed the Rocky Mountains has left Colorado with four major mountain valley parks: North Park, Middle Park, South Park, and the San Luis Valley. The floor of each mountain valley park is the upper surface of thick sedimentary beds supplied by erosion of the mountain ranges. The western edge of the Rocky Mountains is, in part, defined by the Grand Hogback between Rifle and Meeker, where the sedimentary beds were left nearly vertical after the uplift. To the west is the deeply incised Colorado Plateau region. The high plateau is part of the reason the western edge of the Rockies is not a mirror image of the Colorado Front Range.

The principle involved in the uplift of batholiths and mountain ranges can be seen in miniature whenever tree roots grow close to the surface under the asphalt pavement. The ever enlarging root increases pressure to push up the asphalt. As in the scenario outlined above, the asphalt over the root soon becomes highly fractured, making it the target for erosion forces.

Block and sawtooth mountains

Block mountains are formed when twin faults allow a large slab of the crust to be thrust upward as a unit. Fig. 3.13 illustrates a block mountain type uplift where the elevated

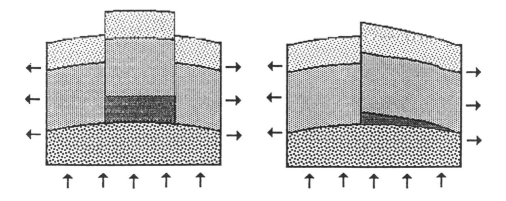

Fig. 3.13 Block mountains. Near parallel faults allow a block of crust to be pushed upward by an increase in magma pressure. Crustal tension opens fault surfaces allowing magma intrusion.

Fig. 3.14 Sawtooth uplift. Similar to block mountain uplifts except that crust is pushed up unevenly creating the cross-sectional appearance of saw teeth. One or more near parallel uplifts may occur side by side.

slab between two vertical faults remains essentially horizontal. Pairs of inclined faults can allow a similar type of uplift with deep seated magma pushing up the crustal block. In most cases the elevated slab will brake free from the adjacent crustal rocks but not elevate uniformly. Fig. 3.14 illustrates a case where one side of a crustal fracture remains attached, while the other side is tilted upward (sawtooth mountains). Some combination of the two block mountain illustrations is more probable than the simplistic formations as illustrated.

Uncompahgre plateau

A massive uplift in western Colorado provides a good illustration of some of the crustal gymnastics we have been discussing. On the north end of the Uncompahgre Plateau is the Colorado National Monument. You can view its vertical red sandstone canyon walls and rock formations from a rimrock drive overlooking the eroded canyons. The rimrock drive can be accessed from Grand Junction, Colorado on the east and Fruita, Colorado on the west. You can see the Colorado National Monument from below along interstate highway I-70, between mile posts 20 and 25, where it appears that a classic batholith has been draped with a red sandstone blanket. The near horizontals sedimentary beds that makeup most of the canyon walls, suddenly dip downward to the North and appear to dive beneath the Grand Valley (Valley of the Colorado River). Evidence of the Uncompahgre uplift can also be found in the much younger Mancos Shale and Mesa Verde formations that form the Bookcliffs along the north edge of the Grand Valley. Their sedimentary beds are tilted upward toward the south, but at a less steep angle, as if they once extended over the Colorado National Monument. Your best view of the tilted Bookcliff formation is from Lands End on Grand Mesa, at the eastern end of the Grand Valley.

The Uncompahgre Plateau extends south to the base of the younger volcanic San Juan Mountains. Geologists tell us the Uncompahgre Plateau was formed when a slab of Ancestral Rockies was uplifted. Faults along the western edge of the plateau allowed the crustal slab to tilt upward similar to our sawtooth example in Fig. 3.14. The Plateau's Precambrian base is dated in billions of years before present, making it much older than any of the sedimentary beds of the Colorado Plateau. The sequence of events started with the Ancestral Rockies being eroded bare before the much younger overlaying sediments were deposited. The Uncompahgre uplift is dated a short 66 million years before present— near the date given for the mass extinction of the dinosaurs. The magma intrusion directly responsible for the Uncompahgre Plateau uplift is not exposed, but volcanism is prevalent in the adjacent areas.

After the uplift, Nature went to work to provide us insight into the crustal structure of its masterpiece. Apparently the Mesa Verde and Mancos Shale formations of the Bookcliffs, mentioned earlier, did extend over at least part of the Plateau. Enough s, at least, to block the drainage of the Colorado and Gunnison River Basins from flowing west as they do now. Instead, their waters were deflected south to carve Unaweep Canyon across the heart of the Uncompahgre Plateau and empty into the Delores River. Unaweep Canyon now exposes the Precambrian crystalline base with granite intrusions. After the softer Mesa Verde and Mancos Shale erosion allowed the Colorado and Gunnison rivers to flow west, Unaweep Canyon was left high and dry as a double-mouthed canyon. Part of the canyon is almost "U" shaped, more typical of glacial erosion as opposed to the "V" shapes of the

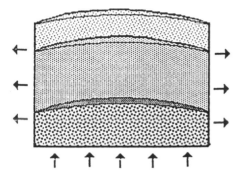

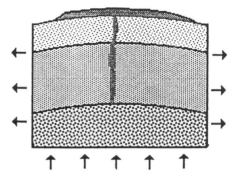

Fig. 3.15 Regional uplift. A slight uplift over a large regional area may not show any of the characteristic surface signatures of a batholith or mountain range intrusion.

Fig. 3.16 Flood basalt. Lava flows up through a dike to spread out on the surface as a flood basalt (called malpais in the southwestern US).

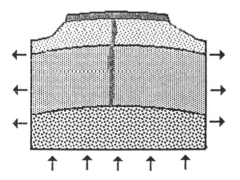

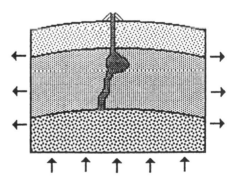

Fig. 3.17 Mesa formed by erosion around a flood basalt. Basalt cap protects the softer sedimentary beds below against erosion. Grand Mesa, in western Colorado, is a basalt-capped mesa nearly 11,000 feet above sea level, and about 5,000 feet above floor of the valley to its west side.

Fig. 3.18 Birth of a volcano.

Black Canyon of the Gunnison River a few miles to the east, or the Royal Gorge of the Arkansas River.

Regional uplift

A large regional area of the crust can, of course, also be pushed up only a few inches or feet without leaving any of the characteristic signatures of a batholith intrusion. The volume of the magma intrusion may even be greater for such a broad area uplift than for the more spectacular batholith or mountain range formations. Fig. 3.15 illustrates the subtle arching from a regional uplift.

It is regional uplifts like the Colorado Plateau and the Great Plains that bring into focus a shortcoming in the theory of plate tectonics. Although there are many local intrusions that provide spectacular crustal gymnastics throughout the Colorado Plateau, there is also a general uplift of the region that is evident by the near horizontal sedimentary beds extending over hundreds of miles. The concept of plate tectonics satisfies compressive mountain building mechanics as a product of two plates in collision. Horizontal forces could even account for overthrust belts discussed later in this chapter under "Classic Negative Quadrant Features." Vertical forces necessary to simply lift and lower large regional areas, however, are delegated in the plate tectonics theory primarily to the thermal expansion and contraction of the earth's interior. This raises the question, "could thermal expansion account for vertical movements measured in thousands of feet that is needed to explain the Great Plains and Colorado Plateau uplifts?" If so, should not they also be relatively warm regions of our planet?

For the Dynamic Axis Theory the relative equilibrium of pressure from higher magma seas in the positive quadrants is achieved when the intrusion reduces the magma pressure to the point where the lithosphere capsule is sufficiently strong to contain the magma. The volume of magma intrusion approximates the volume of change in the geoid presented in chapter 2. Using the Dynamic Axis Theory, the many indications of large sea level changes around the world can be explained either as a product of a crustal elevator platform or by a shift of the geoid. It is reasonable to assume that some combination of both is usually involved. The challenge to geologists is to find clues to determine how much of each was involved on a feature by feature basis. Only some very careful and intense investigation will determine the extent to which the present elevated lands are attributable to regional uplift, or to an axis shift that lowered the geoid.

Volcanic activity

In the examples noted above it has been assumed that the increased magma pressure was stabilized within the crustal shell. Volcanic eruptions are simply an extension of magma intrusions with the magma passing completely through the crust's surface. A vogue explanation is that a heated plume of magma has worked its way up into the crust and formed a magma chamber. From time to time the heated magma expands and is forced up through an orifice or dike fracture where it is discharged either as a lava flow or explosive eruption.

Lava flows spilling out of Kilauea volcano in Hawaii occur when magma is slowly forced up through a crustal vent and is discharged with only a minor eruptive force. As the lava eruptions continue through an orifice, the classic volcanic cone gradually builds around the vent. Thousands witnessed the eruption of Kilauea from their ringside seats with little personal danger sa homes directly in the path of the lava flow were destroyed. Even during relatively dormant periods of Kilauea, the magma stock has remained close enough to the surface to prevent the lava in the vent from freezing sufficiently to form a tight fitting cork. If the vent sealed tight, and the pressure rebuilt, Kilauea could switch to an explosive type volcano. The water and silica content of the magma also plays a vital role in the type of volcanic eruption we can expect.

The massive basalt beds of the Columbia Plateau in the Washington, Oregon, and Idaho were formed by lava flowing freely from dikes, primarily from an area known as the Grande Ronde Volcano Region in southeastern Washington. The beds of basalt were built up by multiple lava flows, some several hundred feet thick and covering thousands of square miles. The magma pressure alternately increased and subsided over a period of some four million years to account for the laminated layers of flood basalts. Among the places where these thick basalt beds are now exposed to our view are the Columbia River Gorge along interstate highway I-84, and the Scablands around the Grand Coulee Dam.

It is significant to note that the volcanic activity in the Cascades, to the west of the Columbia Plateau, apparently subsided during the period of the active volcanic lava flow. The Cascades became active again when the flood lava eruptions ceased (except for the Snake River plains on the extreme eastern edge of the magma flood plain area). This indicates that the magma pressure existed nearly continuously in the region over an extended period, but it could only support one relief valve at a time.

Geologists also note that the dikes exist because the crust was stretched in an east-west direction. In the Dynamic Axis Theory, increased magma pressure responsible for an eruption is also responsible for an increase in crustal tension to open crustal fractures and facilitate dike formation. Fig. 3.16 illustrates the eruption of flood basalt where lava is ejected on a near level plane. The direction of lava flow from any eruption is determined by the ground slope at the point of discharge. The magma flowing from linear vents is simply an extension of intrusions into dike fractures which reach the earth's surface.

Another major flood basalt feature is the Deccan Traps, in west-central India. It formed about 65 million years ago in a series of eruptions that continued over a period of about 500,000 years. Approximately 480,000 cubic miles of magma poured out of the earth. To bring the volume into perspective, consider a 4.6 mile thick layer of flood basalt covering Colorado, or a 395.4 mile high mound the size of Rhode Island.

Richard A. Kerr's article, "Did a Burst of Volcanism Overheat Ancient Earth?" discussed the Ontong Java Plateau (a lava flow twice the size of Alaska and 25 miles thick on the ocean floor near the Solomon Islands—dated at 120 million years before present) that rivals the Deccan Traps in volume of extruded magma. He raised the question of how periods of extensive lava flow could directly and indirectly affect regional and world climates. Flood basalts on land would dissipate the heat by radiation and convection that would raise atmospheric temperatures downwind. The same eruption on the ocean floor would dissipate the heat by convection through the ocean waters and atmosphere, affecting not only temperatures but evaporation and precipitation rates. It is not difficult to visualize how even small eruptions of flood basalts on the ocean floor could alter ocean currents, such as the El Niño. While some heat is lost by radiation from the warmer ocean waters, more heat is available to influence global climate when flood basalts occur on the ocean floor. Scientists at the 1991 spring meeting of the Geophysical Union, held in Baltimore and Pasadena, suggested the superplume eruptions were partly responsible for the lush growth needed to create oil and coal deposits. The Dynamic Axis Theory has no quarrel with some global swings in environment. It does, however, suggest an axis shift that triggered the superplumes, and that climate swings are more often regional than global—regional in the sense of continents and quadrants.

R. S. White and D. P. McKenzie's article "Volcanism at Rifts" (109) discussed underwater flood basalt flows associated with the Iceland Hot Spot in the North Atlantic.

Many buttes and mesas are the result of flood basalt discharging through an orifice and flowing out on near level terrain. Years of erosion eventually leave the high ground as a dominant geologic feature. Grand Mesa in western Colorado, the highest flat top mountain in the world at about 11,000 feet elevation, is capped by a basalt lava flow that protects the softer shales and sandstone beds below from erosion. Grand Mesa stands over 5,000 feet above the Grand Valley, making its western face an impressive sight from any where in the valley. The Flattops northwest of Glenwood Springs, Colorado, and the Table Mountains near Golden, Colorado, are other examples of volcanic lava flows providing an erosion resistant cap.

Although nothing comparable to the Columbia Plateau and Deccan Trap lava flows have been recorded in history, the malpais lava beds south of Grants, New Mexico contain

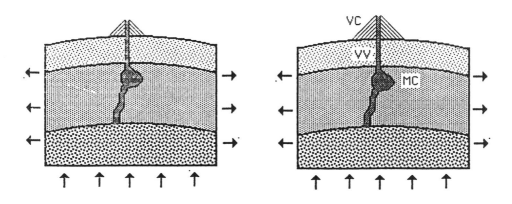

Fig. 3.19 Growing volcanic cone.   **Fig. 3.20** Mature volcano.

VC = volcanic cone; C = caldera or crater; MC = magma chamber; VV = volcanic vent
Magma moves up through the crust into cavities known as magma chambers. As pressure builds in the magma chamber, the magma continues to move up the volcano's vent. The volcano's cone grows and each eruption spills more lava and ash. When the pressure is insufficient to keep magma flowing,the magma freezes and plugs the volcano, until sufficient pressure builds to dislodge the plug.

flows as recent as 1,000 years ago. An arm of the 30 mile lava flow reaches interstate highway I-40 just to the east of Grants. The Crater of the Moon National Monument, in Idaho, features fissures, vents, volcanic cones, and lava flows active from about 15,000 years ago up to about 2,000 years ago.

Paricutín Volcano in Mexico gave scientists a rare opportunity to witness the birth of a new volcano on 20 February 1943. When Paricutín volcano erupted, it literally grew out of a corn field. It grew 33 feet the first day, 550 feet the first week, and 1,000 feet in a year. After 9 years and 12 days, Paricutín suddenly cooled down after reaching a height of 1,353 feet. Future eruptions will undoubtedly add to its classic cone shaped volcanic mountain. Figs. 3.18- 3.20 illustrate the growth of eruptive volcanoes that deposit the bulk of their debris in the immediate vicinity of the vent.

Any weakness in the walls of the crustal orifice vent, that feeds magma up into a volcano crater, allows the magma to expand and form a magma chamber. Medical doctors might compare the magma chamber to a hiatal hernia. Victims of hiatal hernias can appreciate the volcanic like eruptions when gas pressure builds in their hiatus chamber.

Explosive eruptions, such as that of Mount St. Helens receive the highest attention. The lava in Mount St. Helens vent from previous eruptions had solidified to form a tight cork. As pressure from intruding magma built up, the frozen lava cork eventually dislodged. That was the day volcanic debris was thrust thousands of feet into the atmosphere. Since Mount St. Helens ended its 123 year dormant state on 18 May 1980, the

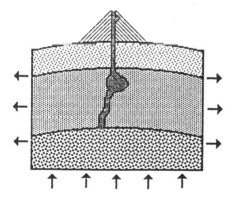 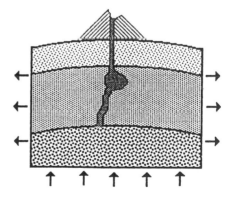

Fig. 3.21 Mount St. Helens before May 18 1980 eruption. Frozen magma in the vent of Mount St. Helens provided a cork that sealed the volcanic cone until tremendous pressure from new intruding magma surge was able to break the seal.

Fig. 3.22 Mount St Helens after May 18 1980 eruption. The frozen plug that sealed Mount St. Helens forced the intruding magma to seek a new outlet on the side of the cone. The eventual eruption was directed toward the weaker north side vents. The eruption blew the top off Mount St. Helens and created an amphitheater shaped crater.

lava dome in its crater has been reformed only to be popped out again on 25 May, 12 June, 22 July, 7 August, and 16-18 October of the same year. Eruptions are continuing at a reduced frequency with no assurance that the mountain is ready for another long nap. The intensity of each eruption depends on the amount of pressure that builds up before the reformed lava plug is released. The lava viscosity, dissolved gases in the magma, and quantity of ground water near its vent are additional factors in the explosiveness of an eruption.

As the volcanic cone grows higher, so does the chance that the frozen magma plug becomes stronger than the side vents that develop in the body of the volcanic cone. When this happens, the next eruption will have a horizontal thrust component that blows the side out of the cone. In the case of Mount St. Helens, much of the eruptive force was directed toward the North to create an amphitheater shaped crater when the volcano erupted. Fig. 3.21 and 3.22 provide before and after illustrations of the eruption of Mount St. Helens that destroyed part of its pre-eruption cone.

Larger volcanoes

As volcanoes go, Mount St. Helens was just a pop gun. When Krakatau (sometimes spelled Krakatao or Krakatoa) erupted on 27 August 1883, a 2,600 foot high island mountain in the Straights of Java all but disappeared from the face of the globe. The article "The Eruption of Krakatau" by Peter Francis and Stephen Self captures the story of the island's destruction. Approximately 36,000 people died, mostly from the tsunami that swept over coastal towns.

In 1815 the much larger Tambora Volcano erupted, also in the East Indies, ejecting some 25 cubic miles of debris and reducing the mountain's height by some 4,200 feet. Henry and Elizabeth Strommel's article "The Year Without a Summer" reminds us that Tambora was credited with or blamed for a dramatic climate change. When it erupted in 1816, its debris was cast high into the atmosphere and stratosphere to cut out some of the sun's warming rays. Killing frosts occurred in New England on 6-11 June, 9 July, and 21 August of that year. Crop losses were extensive, bringing additional hardship to the struggling pioneers.

Mount Mazama's eruption about 7,000 years ago left us with a spectacular 2,000 foot deep lake in its caldera that is probably better known as Crater Lake, Oregon. Scientists have used mini submarines to explore Crater Lake's depths. Their findings are in part comparable to the explorations of the midocean ridges. Strange life forms were discovered around the submarine vents where hot gas and water was still escaping through the crust. Scientists have also made an association between the formation of the world's mineral belts and submarine vents that spew lava and gases out from the earth's interior.

When the driving force behind magma intrusions ceases to act, the magma in the vent freezes solid. If the magma pressure never regains enough thrust to pop the cork, the volcano becomes extinct. Nature's erosion process then takes over, as if she wanted to erase away past geologic history. What remains after years of erosion, however, helps to complete the story on how volcanoes are born and die. Fig. 3.23 illustrates the eventual destruction of the classic volcano by erosion.

Shiprock, in northwestern New Mexico, and Devils Tower in northeastern Wyoming represent the final plugs in ancient volcanoes that permanently sealed their vent. Erosion has removed the relatively softer rock and soil of the volcanic cone to expose the harder igneous and metamorphic stock of the volcanoes vent. Fig. 3.23 does not show the dike formations that are part of the Shiprock geology. As the magma pushed up through the vent, it fractured the earth's crust radially out from the main vent. The resulting fractures were filled with high pressure magma that cooled to form dikes. The erosion process has now exposed dikes that stand above the arid lands of northwestern New Mexico, all pointing toward the parent Shiprock Monument.

Giant calderas

Krakatau, Mount Tambora, and Mount Mazama should be considered as medium size volcanic eruptions in comparison to some of the giant calderas (depression where the surface crust collapsed after the intruding magma subsided) around the world. The Yellowstone Caldera in northwestern Wyoming is one of the giant volcanic eruptions dwarfing the above mentioned volcanoes. Its discharge of pumice and ash was over 6 times greater than that of Tambora, 50 times Krakatau, and over 16,000 times the size of Mount St. Helens.

Yellowstone erupted some 600,000 years ago leaving an elongated caldera over 40 miles in its largest dimension. The area still has considerable seismic activity with its geysers and hot springs. The Hebgen earthquake, just west of Yellowstone in Idaho, attests to the deep seated forces still acting on the region.

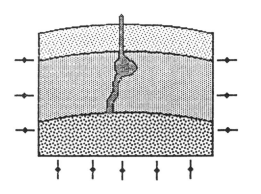

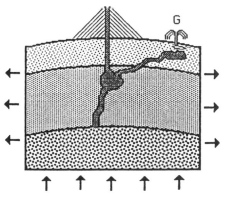

Fig. 3.23 Frozen volcanic vent exposed volcanic intrusion. Shiprock in northwestern New Mexico, and Devils Tower in northeastern Wyoming are classic examples of ancient volcanoes that have surrendered their conic mountain to erosion.

Fig. 3.24 Hot spring geyser associated with erosion. The geyser shown on the right side of the volcano is powered by a magma-heated pocket of water.

Geologists explain the volcanic activity in Yellowstone as a product of a hot spot that resides under the North American Plate. The same hot spot is assumed to be responsible for older volcanic activity to the west in Idaho. As the North American Plate moved west over the hot spot, the active eruptions appeared to move to the east. Because of the size of the Yellowstone Caldera, and since it is a resurgent one (floor domed upward to partially refill the original sunken caldera), it was not recognized as a caldera until fairly recently.

Caldera eruptions in Long Valley, California 700,000 years ago and Valles, New Mexico 1,000,000 years ago are among the more recent of the giant volcanic calderas in the United States. At least 18 calderas, formed 20 to 30 million years ago, have been discovered in the San Juan Mountain Region of Colorado. Remote sensing by satellite has aided in the search for the giant calderas. The satellite perspectives allow geologist to back off and look at the big picture. For more details, see Peter Francis's "Giant Volcanic Calderas."

Geysers and hot springs

As the magma moves close to the surface, the potential of it heating pockets of underground water increases. The heated water pockets form steam pressure that can combine with the normal magma pressures to increase the potential and intensity of a volcanic eruption. It can also simply be the driving force behind hot springs and geysers that may be associated with either active or semi-active volcanic regions. The same hot spot (plume of magma that torched its way up into the crust) that triggered the caldera eruptions mentioned above, fuels the Yellowstone Geysers.

Tectonic plate spreading

Vertical forces responsible for volcanoes are primarily attributable to increased magma pressure. Tension forces on the crust also play a vital role by opening crustal fractures and allowing the magma to work up into the crust. The most active magma intrusions in progress today are along the midocean ridges that cut through all the major ocean basins. One of many articles on the subject of tectonic plate spreading is Jean Francheteau's "The Ocean Crust." As the tectonic plates spread, magma from the interior pushes up to fill the gap. This cycle has been repeated many times, with the intruding magma being cooled rapidly as it is exposed to the ocean waters. The surge of magma forms near vertical seams of magma deposits, as opposed to the near horizontal seams for the multiple layers of flood basalt.

Dating of the vertical seams on both sides of the midocean ridge has produced a mirror image of the forming rocks. Younger rocks are near the ridge with progressively older rocks being found as one moves away from the ridge in both directions. Chronological dating of successive layers is one of many discoveries that has provided strong support for the plate tectonics theory and plate spreading. Fig. 3.25 illustrates a cross section of the magma intrusion that forms the midocean ridge and its rift valley.

The tectonic plate spreading along the midocean ridge can be explained as a product of a positive quadrant. As long as the midocean ridge remains in a positive quadrant over an extended geologic period (multiple axis movements), the increasing great circle circumferences will continue to pull the tectonic plates apart and add new magma to the ocean crust. Interestingly, the entire S-shaped midocean ridges in the North and South Atlantic could be in a positive quadrant at the same time, if the Western Hemisphere's neutral meridional great circle crossed the equator at about 20-30 degrees west longitude (Fig. 3.26).

Thus, we can see that the Dynamic Axis Theory provides an alternative hypothesis to the theory that the upwelling of magma along the midocean ridge creates conveyor belt magma currents that carry tectonic plates in opposite directions. A more detailed discussion of the differences between the conveyor belt theory and the Dynamic Axis Theory is covered near the end of chapter 5.

Classic Negative Quadrant Features

Compressive forces on the crust can explain geologic features, such as synclines, anticlines, deep ocean trenches with their subduction zones, and overthrust belts. Regional subsidence is simply the reverse of regional uplift discussed under "Classic Positive Quadrant Features." Any failure of the crust attempting to bridge the magma gaps formed in the negative quadrants will result in local or regional subsidence as the crust sinks to rest on the lowered magma seas. Any over-correcting by the crust, as it collapses or is driven down into the magma interior like a hydraulic cylinder plunger, creates a local increase in magma pressure. This can explain how local positive pressure is creating volcanic activity along negative quadrant crustal features, such as the subduction zones and overthrust belts

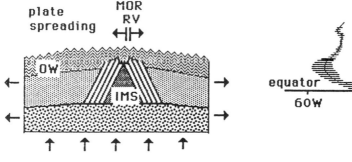 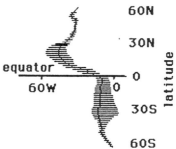

Fig. 3.25 Midocean ridge and rift valley.
OW = ocean waters that cool the magma;
RV = rift valley
IMS = intruding magma seas
MOR = midocean ridge.

Fig. 3.26 S-shaped Mid-Atlantic Ridge. Based on the National Geographic Physical Globe, it appears that the longest transverse faults occur at 25 degrees north latitude and 30 degrees south latitude. Inflection point in S-shape ridge at about 20 degrees west longitude near the equator.

around the Pacific Ocean that makeup the so-called Ring of Fire. One can easily demonstrate the over-correcting of the plate being pushed down into the magma sea: step on a board in soft mud and watch the mud and water erupt up through knotholes in the board (watch your eyes—experience talking).

The most probable sites for compression distortion or failure are areas where the crust is relatively thin. This can be either under the oceans where the crust is naturally thinner, or along active fracture zones that define plate edges. Another possibility is in areas where magma intrusions have torched their way up into the crust during periods when the crust is under a positive quadrant influence (hot spots). Of course, inclined fractures in the crust, both along plate edges or within the tectonic plates, can also facilitate compression (dip slip) failures. Consequently, diagrams with a thinner lithosphere will illustrate the effect of compressive forces in the negative quadrants.

Only the crust that stretches over higher magma seas will be under tension. Compressive forces exist within any plate being driven across either negative or positive magma seas, regardless of the source of the driving force. The driving force we are concerned about is the vector sum of all forces attempting to move the plate. Compression occurs whenever there is resistance to the driving forces. Resistance results from friction between the crust and magma seas, and from friction between plates sliding past each other. Other sources are the inertia of a plate that is in the path of the plate being driven, and the inertia of the leading edge of the plate being driven. When strong driving forces exist along with stiff resistance that prevent free plate movement, the crust will buckle.

Anticlines and synclines

Anticlines and synclines are similar to the folds we see in the hallway rug after junior came in the house too fast and tried to stop. In the lithosphere, a relatively minor collapse of a crustal block can trigger the formation of a series of anticlines and synclines, if the compressive forces are maintained on the crust. Figs. 3.27-3.29 show a sequence of steps of crustal buckling. When a block of crust sags, continued horizontal compression forces cause the crust to increase the syncline buckling (Fig. 3.27). The crustal fold is the same response we expect when we apply compressive forces to the edge of a piece of paper. The crust, restricted in its freedom of sagging by the supporting magma (when the void of a negative quadrant has been filled), will develop an anticline fold as shown in Fig. 3.28. (If we apply compressive forces to a piece of paper resting on a table, it will only buckle up as an anticline fold due to the 100% support provided by the table.) Continual application of crustal compression will increase the sharpness of the fold as shown in Fig. 3.29. The upward buckling of the companion anticline can create additional magma void, which in turn triggers the start of another syncline. Fig. 3.30 illustrates the development of a series of anticline/syncline folds. The anticline/syncline ridges of the Appalachian Mountains are prime examples of crustal folding. When exposed by erosion, the waving pattern of the normally horizontal sedimentary bedding accents the geologic signature of the folds.

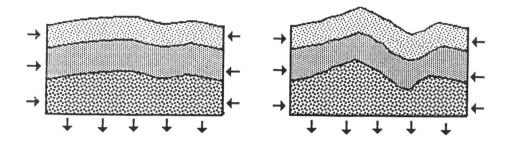

Fig. 3.27 Minor sag in crustal plate creating the start of a syncline

Fig. 3.28 Increased sagging with crustal compression creating a small syncline/anticline fold.

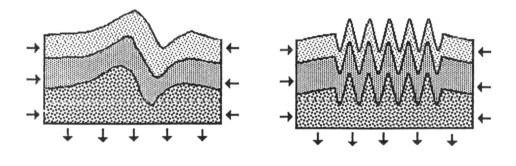

Fig. 3.29 Sharpening syncline/anticline fold.

Fig. 3.30 Series of syncline/anticline folds.

Ocean trenches and subduction zones

The deep ocean trenches represent a major crustal collapse. The naturally thin ocean crust becomes even thinner and weaker as some melting occurs on the crust pushed lower into the magma seas. M. Naïf Toksoz's article "The Subduction of the Lithosphere" offers greater detail of the plate subduction process. Continued compression forces not only create a giant syncline, but one that can rupture the crust to allow the crustal slab on one side of the syncline to slide under the other plate. If the crustal fold and rupture occur along a continental or ocean basin interface, the thinner ocean plate will normally become the subducting plate. Depending on the angle established by the original crustal collapse,

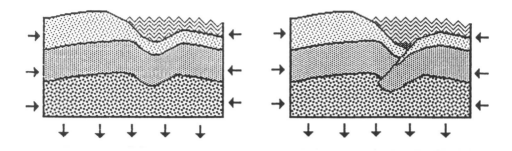

Fig. 3.31 Minor crustal collapse leading to plate subduction. Initial syncline formed when negative quadrant pressure is lowered and magma void occurs.

Fig. 3.32 Major crustal collapse. As the negative pressure deepens, the syncline fold sharpens. Crust ruptures along high stress line in fold.

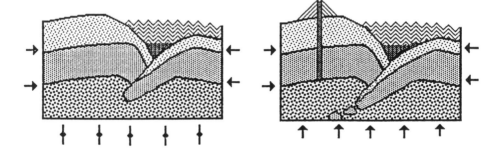

Fig. 3.33 Development of subducting plate. Continued crustal compression forces drives subduction plate into the magma sea.

Fig. 3.34 Overcompensation of subducting plate provides local positive pressure to trigger volcanic activity. Crustal compression forces continue to drive plate deeper into magma sea. Negative magma pressure switches to positive pressure as subduction plate drives deeper into magma sea. Positive magma pressure is released by a series of volcanic eruptions.

the subducting slab will often dive back into the mantle at a fairly steep angle to be remelted. Figs. 3.31-3.33 illustrate the sequence of steps involved in the creation of a deep ocean trench and subduction zone as suggested above. With continuing horizontal compression forces driving the subducting crust deeper into the mantle, as shown in Fig. 3.34, the negative or neutral magma pressure becomes positive. Positive magma pressure created by the plunger action of the subducting plate will cause some vertical lifting of the upper plate. The local positive magma pressure seeks release through crustal openings as volcanic activity.

An alternate scenario for the development of a subduction zone involves the acute diagonal fault as shown in Fig. D 7, Appendix D. The same driving force exists as in the above scenario. With a negative pressure in the underlying magma, the lower plate drops to fill the gap. Since the upper plate cannot support itself, it also rests on the lower plate and pushes it deeper into the magma sea. Continued compression forces drive the lower plate past the upper plate, bending it downward at an angle that extends the subducting plate deep into the magma sea.

The Aleutian and Kuril trenches across the northern Pacific are nearly perpendicular to the Pacific Plate's path, as the latter collides with the Eurasian Plate. A complex system of deep ocean trenches exists in the western Pacific and northeastern Indian Oceans. These reflect plate collisions involving the Pacific, Indo-Australian, Philippine, and Caroline plates. If the tectonic plate moves as a unit, some trenches will have a sizable lateral shear component along with a perpendicular component. Obviously, not all trenches around the Pacific Ocean can be explained as head-on collisions. The Mid-American Trench, along the west coast of Central America results from the collision of the Cocos Plate with the North American and Caribbean plates. The Peru-Chile Trench along the west coast of South America, absorbs the spreading from the Nazca Plate at the East Pacific Rise and the South American Plate moving away from the Mid-Atlantic Ridges.

It is interesting that the Kuril, Japan, Izu, Mariana, New Britain, New Hebrides, and Kermadec trenches in the western Pacific form an S-shaped curve very similar in size and shape to the S-shaped Mid-Atlantic Ridge. These two major features on our planet are approximately 180 degrees apart. The Dynamic Axis Theory calls for crust expansion in the positive quadrants and contraction in the negative quadrants. Is it just a coincidence that axis movements creating a neutral meridional great circle near 30 degrees west and 150 degrees east longitude on the equator, move all the Mid-Atlantic Ridge in positive quadrants, whilst all the above mentioned trenches fall in the negative quadrants?

Overthrust belt

The formation of an overthrust belt is similar to the formation of deep ocean trenches and subduction zones except that the crustal thrust causes one plate to override the other. If magma pressure keeps the lower crustal block from depressing deeper into the magma sea, the overriding crust has nowhere to go except over the top. The weight of the overriding plate will, however, keep it from continuing skyward as a mirror image of a subducting plate. Figs. 3.35-3.38 illustrate a sequence of overthrust belt formation initiated by a syncline fold. Early in plate collision mechanics, the negative pressure is neutralized or reversed. Continued application of compressive forces pushes one plate over the opposing plate that is fully supported by the magma sea.

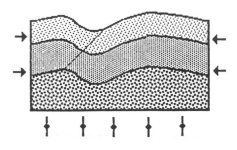

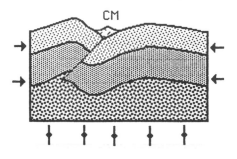

Fig. 3.35 Compressive forces on crustal plate leading to overthrust plate. General crustal collapse eliminates formation of magma void. Compression forces continue to act. High magma seas prevent lower plate from diving into magma sea. Dip angle of fault determines which plate will override the other.

Fig. 3.36 Crustal fracture initiating overthrust. Compression forces drive upper plate along fault line.
CM = crustal moraine formed by upper plate scraping dirt and gravel off the lower plate.

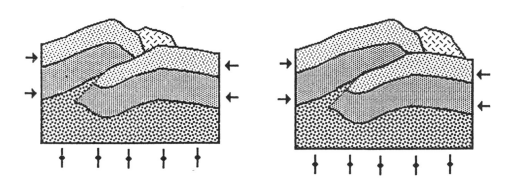

Fig. 3.37 Crustal overthrust continues.   **Fig. 3.38** Crustal overthrust continues.

The acute fault shown in Fig. D 7 (Appendix D) provides an alternate overthrust scenario. Again, the magma seas support the lower plate to prevent subduction.

Some local volcanic activity may also be associated with the formation of the overthrust belt, since the weight of the stacked plates will push crust deeper into the magma sea to create a local positive pressure (Fig. 3.29). Volcanic eruptions will occur, if a crustal fracture or an orifice allows the increased magma pressure to escape.

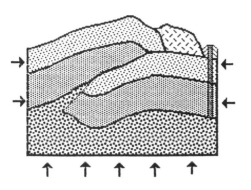

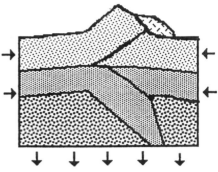

Fig. 3.39 Overthrust belt and volcanic activity. Crustal loading from the two plates pushes them down into magma sea, creating positive magma pressure. Positive magma pressure is released by volcanic eruption.

Fig. 3.40 Combination subduction and over-thrust.

Fig. 3.40 shows how a wedge created by dual faults can create both a subduction zone and overthrust belt when compressive forces drive one plate into another plate like a wedge. This dual fault illustration challenges geologists to find a classic example or to prove that it cannot happen. The collision between the Indo-Australian Plate and the Eurasian Plate that formed the Himalayas would be a good place to search for evidence.

Any differential in the parallel components of forces that act on the crust generatesshear forces. Consider ten people in a line with their arms locked together. If they begin to walk either forward or backward in step, the line remains straight. If each person in the line picks up the pace just a little more than the person on their right, the line can still remain straight, but becomes skewed relative to the original alignment. Each person in the line would feel some shear stress in their locked arms. If the five people on the right began to race ahead of the five on the left, the shear stress between the fifth and sixth person in the line would increase substantially, and possibly cause them to break their arm locks. This is comparable with events along the San Andreas fault. Of course, the shear forces in the above example would be even greater if the five people on the left attempted to go backwards, while the other five continued forward. In either case, some of the shear force is felt at a decreasing magnitude by people on each side of the center.

When we open a phone book on a table, the differential forces between the pages cause each page to slip to differing extents. An end view of the phone book is comparable to the family of transverse fractures in the earth crust next to the midocean ridge. Shear stress exists throughout the crust since tectonic plates are being pushed, pulled, and rotated. The crust is strong enough to withstand many of the differential stresses. Only when the shear stresses coincide with a relative weakness in the crust, will the crust rupture and the seismic needles flutter.

Plate edge faults

Plate edges are the seismically most active regions on the planet. The type and intensity of seismic activity depend on the relative movement of the plates. Along the midocean ridges, where the plates are being pulled or pushed apart, earthquakes occur frequently. Midocean ridge earthquakes, however, are usually of the weaker variety since the crust, weak in tensile strength, cannot accumulate large stresses.

The edges between plates in direct collision experience many large earthquakes as the friction lock and inertia team up to build stress, or as the brittle crust twists and folds. A fault between two plates in collision can rupture, as one or both plates bend into a syncline or anticline folds, when one plate subducts under the other, or as one plate is thrust over the other. New ruptures occur wherever the crust is distorted. Most fault ruptures for subducting or overthrusting plates, however, occur along existing dip slip faults.

The San Andreas fault zone—composed of several near parallel faults—between the Pacific and North American plates is a case in point. It ruptures as strike slip faults that relieve the stress of the Pacific Plate's movement to the north-northwest approximately parallel to the fault zone. The faults rupture in segments along one of the faults where stress exceeds the friction lock. As the North American Plate moves toward the west (away from the Mid-Atlantic Ridge), the friction lock along the San Andreas Fault increases. Additional factors in the friction locks are the bends in the fault alignment. Before the Pacific Plate can move north-northwest relative to the North American Plate, it must also overcome the friction lock. The latter result from spurs on the North American Plate that jet out into the path of the Pacific Plate. Appendix D presents a more detailed analysis of plate movements effects on earthquake intensity.

We must also accommodate the westerly movement of the North American Plate. Several of the earthquakes along the San Andreas involve dip slip faults that relieve the collision forces and account for uplifts of the coastal range. To the north along Oregon, Washington, and Canada, the westerly movement of the North American Plate overrides the Juan de Fuca Plate. To the south along Central America, the southern tip of the North American Plate overrides the Cocos Plate. Since relative plate movement seldom provides pure collision or pure shear, each earthquake usually has both a strike slip and dip slip component.

Foreign terranes

When two vehicles collide, either head on or by a glancing encounter, parts of one vehicle can brake off and become imbedded in the other vehicle. If two tectonic plates collide, something similar happens. When a block of the earth's crust is sheared off its parent plate and becomes attached to a second plate, geologists refer to the block as a foreign terrane. When a tectonic plate is subducted under another plate, islands and subterranean mountains may be sheared off its top to become a foreign terrane. Terranes may also be stripped from the edge of two plates sliding past each other. The line that marks the boundary between the Pacific and North American plates is far from a straight line. Spurs or blocks of one plate that projects into the path of the other plate are subject to being sheared off and carried along with the opposite moving plate. In time it will probably become fused to the second plate as a foreign terrane.

Six terranes parked along the coast of southeastern Alaska are still being molded and fused to the North American Plate. Discontinuity across the faults that join the terranes, their fossil imprints, and paleomagnetic signatures indicate to geologists that they are not native to the area. The six terranes caught between the Pacific and North American plates have been distorted and turned clockwise. Heat and pressure generated by the collision has metamorphosed much of the rocks in the area.

It is interesting, however, that the East Pacific Ridge does not indicate a north-south displacement between the Pacific Plate and the Nazca or Cocos plates. One would expect this according to the prediction that Los Angeles will eventually move north to the San Francisco area. The East Pacific Ridge also does not have the same characteristic long transverse faults as are found along the Mid-Atlantic Ridge. It appears that much of the shear stress was absorbed by a 700 mile offset in the Eltanin Fracture Zone about midway between New Zealand and Cape Horn, Africa.

Geologists use the paleomagnetic and fossil data as clues to the origin of the terranes. Paleomagnetic data, however, can distort the clues in many ways. Due to the instability of magnetic poles geodesists have to develop isomagnetic charts that show the declination and inclination as a function of time. Isometric charts are then used by surveyors to correct the magnetic signatures. Local anomalies play a big role in plotting isometric charts. Unfortunately, isometric charts for periods of geologic history are not available and can not be accurately determined at this time. Consequently, past movements of the magnetic poles and the influence of a local magnetic anomaly can significantly distort the paleomagnetic data. Chapter 4 includes additional discussions about paleomagnetic data.

Transverse faults

The greatest concentration of shear fracturing is found deep within the ocean basins associated with the midocean ridges. It was not until oceanographers began to produce maps, like the National Geographic Physical Globe, showing the ocean floor relief, that the family of transverse fracture zones became apparent. The parallel fractures extend perpendicular out from the midocean ridge. The longest fractures appear to be at about 30 degrees latitude, decreasing in length both towards the pole and equator. The family of transverse faults is also associated with offsets in the midocean ridge.

Inspection of Table B-2 (Appendix B) reveals the change in great circle length for a one-degree axis shift and provides a clue to the force necessary to fracture the crust in this fashion. The transverse fracture zones in the Atlantic are consistent with the crustal tailoring required to accommodate an axis shift that moves the neutral meridional great circle intersecting with the equator at about 20 degrees west longitude. The fact that the maximum fault length occurs near 30 degrees, as opposed to 45 degrees, can be explained by including the influence of changes in the earth's rate of rotation. Great circle half-lengths based solely on changes in the spin rate are greatest at the equator and decrease toward the poles.

Why do maps of the worlds largest and most frequent earthquakes not include these massive fracture zones? It is in part because the fractures are nearly straight and in agreement with the changes in great circle length. The friction lock is not nearly as strong as for plate edge faults because it is straight (no spurs as along the San Andreas Fault) and parallel to the active crustal forces (no major force is being applied perpendicular to the fault).

Other faults

Regardless of the driving force behind tectonic plates, the forces are distributed throughout the plate. This means that the forces must be transferred across existing fractures within the plate—fractures that can rupture when the conditions are just right. When Nebraska is pushed westward by Iowa and it meets a fault in Wyoming, the force applied by Nebraska will either be transferred across the fault or cause a rupture of the fault. The New Madrid fault zone running north-south near the Mississippi River is an excellent example of how propelling the North American Plate westward can create large earthquakes within the tectonic plate. At the same time, much of the push supplied by the eastern half of the North American Plate is transferred across the New Madrid Fault Zone without causing an earthquake.

4   AXIS WOBBLE AND AXIS DRIFT

While the flat earth theory was in vogue, it was impossible to consider an axis wobble. A flat earth required the ingenuity of Hollywood's best special effects to account for the movement of stars and planets across the sky. The discovery that the earth was an orb spinning on its axis as it sailed around the sun both simplified and complicated astronomy. Since then, the closer astronomers look at our solar system and the universe, the more subtle deviations they find in the dynamic motions of each body.

Instability of the Earth's Axis

Theories that include the concept of axis movements have often been advanced. However, they were unable to gain general acceptance as the cause of major geologic phenomena. When evidence emerged of ice caps far from the present poles, an obvious speculation was that the poles must have occupied a position close to the ancestral ice cap centers. Attempts to explain coal deposits in Arctic and Antarctica regions prompted similar speculation of axis movements that brought their regions much closer to the equator.

In arguments against the concept of pole movements, it was quickly pointed out that the spinning earth acts like a giant gyroscope with the inertia of the equatorial bulge maintaining the spin axis. Although very minor axis movements would be possible, the large axis shifts required to account for ice caps in India, Africa, and South America would be impossible because of the stabilizing affect of the equatorial bulge. This, we now know, is based on an erroneous assumption that the earth's crust is rigid enough to maintain its ellipsoidal planet shape, with the poles and equator remaining in the same position on the orb.

Plate tectonics offered a new explanation for evidence of glaciers in the tropics and tropical rain forests (fuel for coal deposits) in polar regions by simply transporting the tectonic plates into the appropriate regional climates. Plate movements, however, fall short of accounting for the sudden end of the last ice age. An ice cap over two miles thick melted over a period of six to eight thousand years. A few thousand years of continental drift since the peak of the Ice Age are insufficient time to move Canada away from the poles, even if the plates raced at a velocity ten times today's rate. There is a second problem in assuming that plate tectonics provides the trigger for climate change. How could both Canada and Europe be positioned near the North Pole poles at the same time?

In developing a short Dynamic Axis Theory scenario to explain the origin and effect of axis movements, we need to start with our planet spinning in dynamic balance. When weight shifts on and within the earth, it disrupts the spin balance and causes the axis to wobble. The axis, in turn, drifts toward a position of dynamic balance in an effort to dampen an axis wobble. In chapter 2 we discussed how the ocean waters respond to axis movements. The realignment of ocean waters creates an additional weight shift that triggers further axis responses. Chapter 3 addressed how the magma seas respond to axis movements to produce even more weight shifts. Each response and counter response represent the planets attempt to reestablish spin stability. As long as the planet is not in perfect balance, some axis wobble and axis drift will remain. When the magma pressure that results from axis movements is reduced to a point where the crust can resist further distortion, the crustal rupture will begin its healing process. Another cycle begins when the new weight shifts cause the axis to wobble and drift. Theoretically, incremental pole movements could eventually allow the poles to occupy any point on the globe without the necessity of overcoming the stabilizing equatorial bulge—the bulge simply moves with the axis shift. Each new episode of axis movements represents a fresh start—but we are getting ahead of the supporting evidence.

Euler, Kelvin, and Chandler

In 1765, Leonhard Euler, a Swiss mathematician, presented his findings that a spinning rigid body will wobble if rotation is not about the body's axis of symmetry. Considering Euler's idea, the earth would have a 305 day wobble cycle, if it was a rigid body.

Kelvin, in 1862, discussed the possible influence of seasonal shifts in air masses on the stability of the rotational axis. He examined how they would result in an annual wobble cycle of the earth's axis relative to the body of the planet. In 1891, Seth Carlo Chandler, an American geodesist and astronomer, and his colleagues provided the first concrete evidence that the earth's axis is not rigidly fixed. Based on observation stations about 180 degrees apart, in Berlin and Honolulu, he confirmed that latitude changes of as much as 0.3 seconds of arc do occur. It may have been a stroke of luck that the observations were made when the wobble was large enough to be easily identified. As we will see later, the amplitude of the latitude variation has a cyclic pattern. If the observations had been made during a minimum wobble period when the wobble pattern is less definite, Chandler's project could have been put on hold or cut off completely.

Although an annual wobble was known, or at least suspected, the 14-month wobble cycle discovered by Chandler's analysis of the data was a surprise. This cycle is usually referred to as the Chandler wobble, but occasionally also as the free Eulerian precession, even though it does not fit Euler's computed cycle. The axis wobble motion causes the pole to trace an irregular circular path over the wobble cycle of about 14 months. Some have suggested that Euler's predicted axis wobble could be triggered by a strong shock, such as an 8.0 magnitude earthquake or a large asteroid striking the earth. However, the laws governing dynamics state that without continual agitation the wobble would naturally dampen to zero like a harmonic motion of a vibrating bell.

It is convenient to discuss polar motion with the North Pole's path, but we must always remember that the South Poles' motion trace is a near mirror image of the North Pole trace. It is also helpful to consider the pole positions as temporarily fixed at selected instantaneous positions so that we can analyze the changes resulting from any real or theoretical pole movements.

The Chandler wobble involves the physical movement of the earth's axis of rotation with respect to the body of the earth, but not the slippage of the crust over the mantle. Also, the Chandler wobble is distinct from precession and nutation (wobble of the planet as a unit relative to the solar system) where the projected axis describes the conic rotational motion in space as a product of lunar and solar gravitational attraction. The ancient Greeks discovered the recession of the planet's axis. Over a period of about 25,600 years, the earth's axis scribes a conic surface as moon, sun, and planets tug on the earth during their travels through space. Nutation adds 18.6 year cycle perturbations, involving axis wobble or nodding, which are superimposed on the larger precession cycle. James Bradley, a British astronomer, discovered the nutation phase in the axis wobble in 1748.

Since the term wobble is used with precession and nutation as well as the Chandler wobble, it is necessary to be specific about which wobble is being discussed. Unless otherwise specified during our discussion of the Dynamic Axis Theory, the terms "wobble" and "polar motion" will refer to the Chandler type of axis wobble.

Pole monitoring

Every since Chandler's discovery, scientists have been tracking the pole motion as it moves about the body of the earth. The International Latitude Service (ILS) set up a five station network in 1899 to measure and track the pole motion. The Bureau International de l'Heure (BIH), and International Polar Motion Service (IPMS) also published pole movement data. Each service bases their computations on a different combination of observation data, which gives very similar but slightly different results.

The Pole trace data in Table B 3 (Appendix B) furnish a sample of the published pole coordinates for the period from 1899.806-1901.889 as provided by IPMS in Mizusawa, Japan. The coordinates are in arc seconds from the horizontal reference system pole. Positive [x] is measured along Greenwich, while positive [y] is measured along 90 degrees west longitude. The dates are expressed as a numeric fraction of the year. The observations listed are for about the 20th of each month.

By plotting the two coordinates as independent values, we can graphically see their oscillations. Figs. 4.1- 4.4 represent plots for both the [x] and [y] coordinate values covering the period from 1900-1988. The variation in the cycle amplitude relates to the degree of phase agreement between the annual and 14-month components. The apparent phase shift between the [x] and [y] component plots are simply the 90 degree lag time between a maximum [x] and maximum [y] for the circular trace.

Based on the pole motion data collected since 1900, we now examine two major components of the axis wobble: Chandler's discovery with a 435 ± 2.2 day cycle (we will use 14.29-month cycle), and the annual wobble with its 12-month cycle. Both cycle lengths are averages—much like each season where the actual weather seldom fits the calendar. The average amplitude of the Chandler wobble component between 1900-1988 was about 0.33 seconds of arc (diameter of pole trace). The average annual wobble amplitude for the same period was about 0.14 seconds of arc (BIH gives average annual wobble for period from

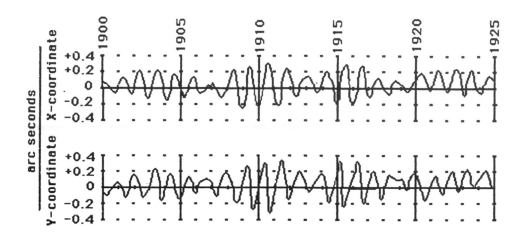

Fig. 4.1 Oscillations in "x" and "y" coordinates of Chandler wobble 1900-1925.

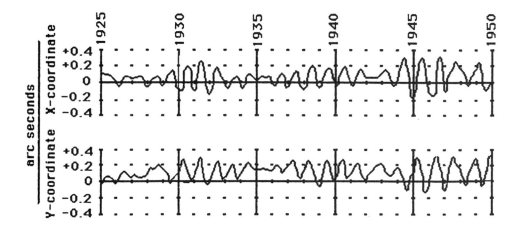

Fig. 4.2 Oscillations in "x" and "y" coordinates of Chandler wobble 1925-1950.

1956-1970 as about 0.18 seconds of arc). The 12- and 14.29-month cycle components both have sizable swings in amplitude from their averages. The two component cycles of 12- and 14-months are in phase (additive) about every seven years when the diameter of the resulting circular traces reach their maximums. At the 3.5 year midpoint in the combined cycles, the two components are out of phase (subtractive), resulting in a much smaller circular trace.

To determine the polar motion contributions by the 12- and 14-month cycles (using Figs. 4.5- 4.8) several assumptions were made, including:

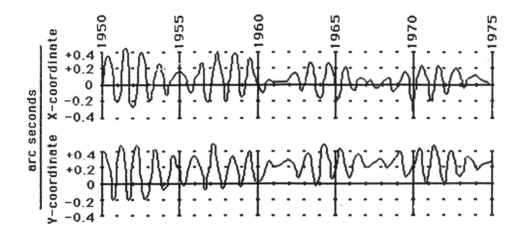

Fig. 4.3 Oscillations in "x" and "y" coordinates of Chandler wobble 1950-1975.

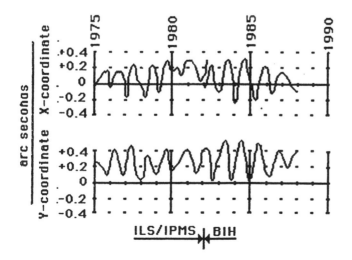

Fig. 4.4 Oscillations in "x" and "y" coordinates of Chandler wobble 1975-1988.

1. Only the 12- and 14-month cycles are present.
2. The contribution of each cycle remains constant.
3. Maximum wobble occurs when the 12- and 14-month cycles are in phase.
4. Minimum wobble occurs when the 12- and 14-month cycles are out of phase.
5. The average maximum wobble, less the average minimum wobble, equals twice the 12-month cycle wobble contribution. Using Figs. 4.5-4.8,

   (Σ peaks - Σ valleys) / number of cycles = twice the annual average.
6. The 14-month contribution equals the average maximum wobble, less the 12-month wobble, as determined under (4) above.

Wobble phase data are provided in Figs. 4.5-4.8. The average wobble (circle trace) diameters were computed and plotted for each 14-month period (arbitrarily starting with the first IPMS record) giving the combined 12- and 14.29-month wobble cycles. The peaks represent periods when the two cycle components are close to, but not necessarily in absolute phase. The valleys are periods when the two cycles are nearly out-of-phase.

Note that two major peaks (maximums) occurred in about 1909 and 1951 (42 years apart), and the amplitude of the intermediate peaks decreases away from the two major peaks. The smallest peak falls near the midpoint between the two largest peaks. The variation in magnitude of the maximums results because the Chandler wobble has a 14.29-month (435 day) cycle, instead of the exact 14-month cycle needed to be in phase every 7 years. A combination of 12- and 14.4- month cycles would be in complete phase agreement every 42 years. While 14.4 months is just outside the range given for the Chandler wobble (435° 2.2-day cycle), a phase shift noted later could account for the difference. Soon we should have the data to confirm, reject, or indicate limitations on the repeatability of the indicated 42-year cycle.

In the remaining section the terms "Chandler wobble", without further qualifications, and "axis wobble" refer to the sum of the two cycle components (12- and 14.29-month). The term "14-month" may also be used in place of the more precise "14.29-month" cycle.

Fig. 4.9 shows a trace of the poles for the period from 1951.056-1952.222 when the Chandler and annual wobble were close to being in phase. A second trace is also shown for the period from 1940.056-1941.222, when the two cycles are close to out of phase. The 1940 pole trace selected for this illustration does not represent the minimum. When the wobble is small (1925-1928, 1962, and 1967-1968), the pole path does not always trace a circle. It is possible that the magnitude of the wobble approaches the noise level of the data collected during that period.

Another interesting aspect about the axis wobble variation is the phase shift that occurred during the (1925-1928) period when the 12- and 14.29-month wobble cycles were out of phase. The timing for peaks and valleys, after the period of minimum wobble, nearly reversed.

Axis drift is apparent in Fig. 4.10 which shows pole traces for the periods 1903.056-1904.222, and 1977.056-1978.222. Assume that the centroid of each circular trace represents the pole if the earth was spinning true at that point. We will note that the pole trace for the period beginning on 1903.056 circles the horizontal reference system pole. However, the trace for the period beginning on 1977.056 is offset from the horizontal reference system pole. This is because of the long term drift in the centroid of individual

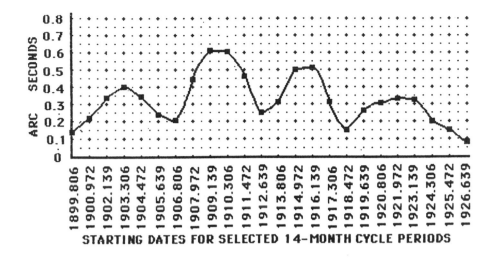

Fig. 4.5 Average diameters for Chandler wobble 114-month cycles (1899-1926).

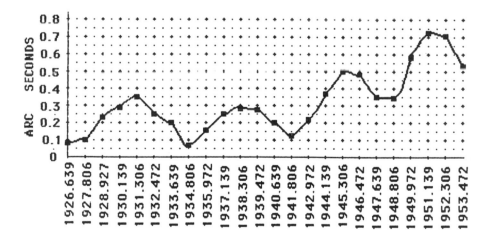

Fig. 4.6 Average diameters for Chandler wobble 14-month cycles (1926-1953).

pole trace plots. The drift can also be detected in Fig. 4.1-4.4 as the peaks and valleys of the [y] component oscillations shift toward the positive. From the available data, the axis drift is far from a uniform movement in both time and direction.

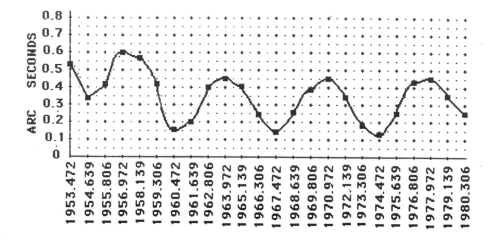

Fig. 4.7 Average diameters for Chandler wobble 14-month cycles (1953-1980).

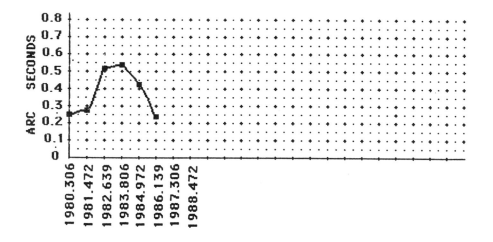

Fig. 4.8 Average diameters for Chandler wobble 14-month cycles (1980-1988).

The Data accuracy for the determination of pole positions from 1900-1988 is limited by several variables that introduce relative errors. The number of locations of observation stations used to track the pole movement did not remain constant over the years—thus changing the reference base. Periodically, the selected zenith stars that were used for the observations had to be changed when they moved away from the observation station

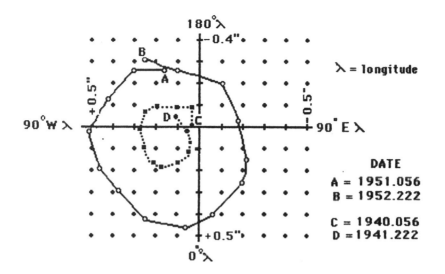

Fig. 4.9 Comparison of Chandler wobbles in phase (1951.056-1952.222) and out of phase (1940.056-1941.2220.

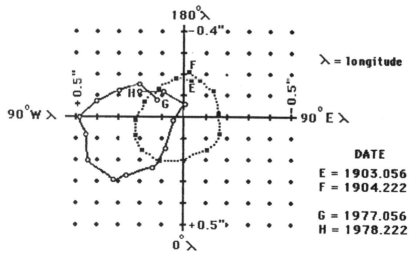

Fig. 4.10 Chandler wobble and axis drift between 1903 and 1977.

zenith. Since the data involves optical observation of stars, clouds cause incomplete data sets for stations during overcast periods. At least one observation station has been physically moved. The observation stations are all located on tectonic plates. This makes it more difficult to separate the contribution made by pole movements from tectonic plate movement. Survey instrumentation and techniques are continually being improved to give a more accurate recording. The tables furnished by IPMS are expected to contain errors as large as 0.1 second of arc. Such errors are undoubtedly responsible for much of the irregularity found in Figs. 4.1-4.4, 4.9, and 4.10. The influence of the random errors, however, is reduced in the cycle plots of Figs. 4.5-4.8 because of the averaging of observations over a 14-month time span. A switch in the available source data from ILS/IPMS to BIH, starting in 1982, is evident by the [x] coordinate plot break in Fig. 4.4.

Higher precision data are needed to eliminate the errors introduced by tectonic plate movement. The relative and absolute movement of the observation stations must be known, so their influence can be removed mathematically in the computation. Alternatively, the reference base for making the observations must be completely outside the planet, so that the data will be free of any error bias introduced by plate movements. Geodesists are utilizing Very-Long-Baseline-Interferometry (VLBI) techniques to establish high precision baseline measurements free from errors introduced by crustal gymnastics. Two or more widely spaced antennas intercept radio signals emitted by quasars at the outer limit of the known universe. The only requirement is that all antennas can point on the quasar without obstructions. Fortunately, clouds are no obstacle for the radio waves. By measuring the difference in arrival time, and using precision atomic clocks, the travel time is converted to a vector distance. Three properly positioned quasars provide three critical vectors that can be used to compute the baseline length.

Baseline lengths that span the ocean are determined within a few centimeters. The quasars, at a distance of billions of light-years, provide fixed references—unlike stars that are close enough for surveyors and astronomers to detect relative movements over time.

Additional cycle contributions

It has also been suggested that the Chandler wobble has two components about 7.3 days apart. If the second cycle is 7.3 days out of phase with the primary cycle, these two components would be in phase about every 50 years. A second possibility is that the indicated blip on the wobble amplitude graph is an echo (aftershock) that will remain close to the primary 14.29-month cycle. Scientists at the Jet Propulsion Laboratory in Pasadena, California and the Atmospheric and Environmental Research in Cambridge, Massachusetts have seen indications of a wobble varying between 2.5-24 inches in diameter, with a cycle period from 2-9 weeks.

Better explanations for several geodetic and geologic questions await the more precise pole motion data. VLBI provides geodesists a tool to add an order of accuracy to their polar motion determinations and resolves many pole motion questions. Geologists and geophysicists need a worldwide network with hundreds of baselines to tie down the relative movements of tectonic plates. As the density of VLBI stations increases, so will the accuracy of existing survey networks that are adjusted to the more precise VLBI network. Conventional surveying techniques, including Global Positioning Satellites (GPS), will provide excellent results for a higher density network tied to the VLBI network. Since the plates are not stationary, time becomes a critical fourth dimension of precision surveys. Repeated baseline measurements that cross tectonic plate edges are needed to refine the records of relative plate movements.

In a closely related application, geodesists are using VLBI data to determine variations in the earth's rate of rotation. Regular cyclic variations affecting the length-of-day over 6-month, 1-month, and 2-week periods have been detected. Computer modeling of the atmosphere's and hydrosphere's cell movements should provide some correlation with these short term spin variations. Longer term variations in the spin rate should provide clues about other shifting masses on and within the earth—especially the river erosion that shifts weights toward or away from the equator.

Since the forces responsible for axis wobble originate from several sources (i.e., erosion, atmospheric pressures, hydrologic cycles, etc.), each produces their specific oscillations. Some of the forces will introduce a near regular cycle period that combines as alternating additive and subtractive components. Random influences on the wobble cycle are also expected. The random influences make the resolution of the puzzle more difficult.

Extrapolating the axis wobble, as determined since 1900, into the past or future will always be suspect. Even an accurate record of pole movements determined over a period of a few thousand years would not necessarily be representative of all geologic history. Evidence of pole motion surges will be discussed in chapter 7.

Axis Movements and Paleomagnetic Signatures

Scientists look for clues to the past where ever they can find them. One of the most fruitful sources are the rocks themselves.

Paleomagnetic studies

When a rock (with magnetic properties) originally forms, the direction and strength of the existing magnetic field are frozen in the rock. Geologists use these magnetic signatures as an indicator of crustal movements that it occurred over geologic history. A major problem relating to paleomagnetic data is that it is referenced to the magnetic pole instead of the earth's spin axis. Unfortunately, the magnetic pole drifts about our horizontal reference system pole—altering the reference base. The present north magnetic pole (at 73°N, 100°W) is about 17 degrees south of the North Pole, whereas the south magnetic pole (at 71°S, 148°E) is about 19 degrees north of the South Pole. At least that is where it was shown on the particular map I used. As we can see, the magnetic poles are not diametrically opposite or symmetric with the body of the earth.

Magnetic pole movements are determined by regularly repeated magnetic surveys used to establish and update isomagnetic charts. The charts show the ever changing magnetic declination (horizontal angle from true north) and inclination (vertical angle below level). The strength of the declination and inclination force components of the earth's magnetic field also vary as a function of time. Surveyors use the isomagnetic charts to correct compass readings at each point within the chart. Knowledge of magnetic pole movements makes the charts dynamic by extending their useful life. Long term magnetic pole movements over geologic ages, however, are not as predictable. The potential for errors derived from accepting the paleomagnetic signatures locked in rocks are illustrated by observations made in London:

Year	Declination
1580	11 degrees east of north (indicating the magnetic pole was off the north coast of USSR)
1812	24 degrees west of north (indicating the magnetic North Pole was far south of its present position)
1947	10 degrees west of north, toward northern Canada

(source - Colliers Encyclopedia)

The declination values noted above would be even larger for an observation station north of London and closer to the magnetic pole. A corresponding compass pointing from New York or Central Russia could have little or no declination variation for the same period, if the magnetic pole for 1580 and 1812 were in alignment with the 1947 location. The effect of magnetic pole movements is most pronounced for observations taken close to the magnetic pole and perpendicular to the path of the magnetic pole movement.

Besides the variations in declination angles between the astronomically determined north (horizontal reverence system meridian) and the magnetic pole, magnetic pointing is also affected by local magnetic fields caused by pockets of strongly magnetized rocks. The magnetic surveys conducted today identify these anomalies, so that corrective measures can be taken. Magnetic anomalies that existed at the time paleomagnetic signatures were frozen in the ancient rocks they present a greater challenge to identify and resolve. If they are not recognized as an anomaly, they can lead to erroneous analysis and conclusions. Consequently, reconstructing plate translations and rotation, based on paleomagnetic signatures becomes very tricky at best. Since it is the best data available, we have little choice but to use it.

Changes in reference base have affected our analysis of pole movements. Over the years, several plots were developed that show a path of the North Pole moving erratically over the earth's surface. Before plate tectonics, the plates were considered fixed and all indications of change from paleomagnetic signatures were assigned to pole movements. By combining the movements of the tectonic plates, the magnetic pole, and the axis, we can imagine how large errors can creep into the determination of ancestral pole paths. Because of the uncertainties of paleomagnetic data, it is easy to understand why two scientists, working with different data sets, can generate plots that appear to have little relationship to each other.

The theory of plate tectonics terminated much of the speculation on pole movements during a period when it was assumed that the plates, rather than the poles were moving. All paleomagnetic signature variations were assigned to plate movement. A closer look at the available data, however, has renewed scientists' efforts to define both pole and plate movements. Armed with more paleomagnetic observations, better dating techniques, and awareness of plate tectonics, scientists are now achieving about 80 per cent agreement when their resulting ancestral pole path plots are compared.

Recent pole movement studies, based on paleomagnetic data, have the poles moving either faster than the plates or slower. Geodetic observations, available only since Chandler's discovery, show that the North Pole has drifted about 30 feet toward New Brunswick, Canada since 1900. This is about twice the rate presently assigned to plate motion. The most cited Livermore study on polar movements claims only 5 degrees of wander occurred during the past 90 million years (about 7% of the present rate of about 0.0032 arc seconds per year or one degree per 1.25 million years). For the period of 100-200 million years ago, their data indicated an axis shift between 17-19 degrees (about 23% of present rate). Based on his experience in refining calculations for the younger periods, Mr. Livermore feels the 17-19 degree estimates may even be too large.

When Richard Gordon, of Northwestern University, teamed up with Roy Livermore to rework the original data, they found evidence for a shift of about 14 degrees occurring some 70-100 million years ago. An averaging of 14 degrees over 30 million years is about half the present drift velocity. To place this in perspective, the present rate of drift of one-degree per 1.25 million years would account for 52 degrees of pole meandering over a 65 million year period—if the movement was steady and in a straight line. This, of course, does not take into account the possibility of axis surges that are discussed in greater detail in subsequent chapters. It also does not consider the possibility of a serpentine path of the poles over the same period.

The Livermore group recognized the complexity of separating plate and pole movements. First they used the sea floor spreading records to reconstruct the erratic wanderings of the plates. They expanded on previous studies by adding data from the huge Pacific Plate to data from the North American and Eurasian plates. Next they located the pole relative to the plates. The pole positions, as determined by dated paleomagnetic signatures corrected for plate movements, still had to be compared to a fixed reference plane. The Livermore group choose the hot spots as their reference. The end product was a plot of the pole path (assuming the identity of the magnetic and North Pole).

The stability of hot spots, however, is suspect. Clement Chase, of the University of Arizona, addressed the uncertainties surrounding the use of hot spots as a plane of reference. He set out to measure the relative movement of individual hot spots in an effort to determine their stability. His data indicated that the hot spots were possibly moving at up to 2.5 centimeters per year. This compares to 5 centimeters per year for typical tectonic plate movement. He also determined that the Northern Pacific hot spots are moving north, the Southern Pacific hot spots are moving south, and the Indian Ocean hot spots are moving northwest. Chase found evidence of unsteady and perhaps random motion of the hot spots over tens of millions of years.

These regional variations in hot spot movements, noted by Chase, are consistent with the idea of magma current patterns being similar to those of atmospheric current patterns as discussed in chapter 5. The present north moving magma current in the Pacific would indicate the hot spots are under the influence of a magma low (counterclockwise rotating cell) to the west of Hawaii. A magma high (clockwise rotating cell) to the east of Hawaii would also provide a northern drag. The concept of magma tornadoes raises more questions. Would they be the product of violent magma circulation when warm and cold magma currents collide? Is the mechanics of hot spots simply a product of local radioactive decay sources in the earth's core that could occur any place within a high or low magma cell? If so, their path would be a function of the magma's circulating pattern at the heat source in the outer core. If they are driven by the upper magma currents, the unsteady or random motion over time will occur as they respond to the magma cell motion.

If the hot spot under Hawaii is moving north at 2.5 centimeters per year, and the

Pacific Plate is moving north-northwest at 5 centimeters per year, the apparent movement of the hot spot is to the Southeast at just less than 2.5 centimeters per year (relative to the crust). If the Pacific Plate movement was determined by assuming a fixed hot spot under Hawaii, the speed of the Pacific Plate would be underestimated by about 50 percent. If Chase's study stands up, hot spots lose much of their value as reference points for determining plate and pole movements. The data analysis based on the assumption of fixed hot spots would be distorted. Adding hot spot movements to the meandering of the magnetic pole creates a tremendous challenge for scientists.

When Gordon and Livermore altered their procedures slightly and reviewed the earlier Livermore data, they found little or no shifting of the outer shell over the past 65 million years. Livermore's determination of a 5-degrees shift over the past 90 million years is presented as being nearly stable. From our discussion of the science fiction scenario in chapter 1, geoid changes in chapter 2, as well as magma sea responses and great circle length variations in chapter 3, even a 5-degree axis movement takes on a whole new meaning that apparently was not considered in the Livermore studies.

Livermore and Gordon's findings are in overall agreement with an earlier study by Gordon and Donna Jurdy (also of Northwestern University). Gordon earlier had found an accelerated shift some 90-80 million years ago, during the late Cretaceous. That study was based only on Pacific paleomagnetic data.

Jean Andrews of the Lamont-Doherty Geological Observatory reported rapid pole movement during the late Cretaceous. Vincent Courtillot and Jean Besse (of the Earth Physics Institute in Paris) found both rapid pole movement and extensive volcanism some 110-30 million years ago. Exact timing of the rapid axis movement seems to present a greater challenge than its magnitude. It would make an interesting discovery if the date for rapid plate movement could be pinned down to the Cretaceous-Tertiary boundary and the demise of the dinosaurs. Chapter 6 discusses past cataclysmic events.

In a similar .study, Vincent Courtillot and Jean Besse calculated what they call an Apparent Pole Wonder Path (APWP). They used paleomagnetic data from the continent of Africa, reinforced by paleomagnetic data from the other major tectonic plates. Courtillot and Besse discussed their ideas in an article "Magnetic Field Reversals, Polar Wander, and Core-Mantle Coupling." They refer to their computed data, which covers the past 200 million years, as synthetic paleomagnetic. For the second step in their effort, they used a synthetic path of the pole that Morgan developed from a model assuming the hot spots are fixed. They refer to the new computed pole path as the True Polar Wander (TPW).

Their True Pole Wander over the past two hundred million years has several interesting features, including:

Date MYBP*	Pole position**	For present geodetic reference system
200	(65°N, 110°E)	-axis drifted to the northeast
170-110	80°N, 130°E)	-pole remained still for 60 million years
		-pole path took a left turn
40	(80°N, 20°W)	-pole path took a right turn
30	(80°N, 60°W)	-pole path made a hairpin turn
present	(90°N,___)	

* million years before present.
** latitude and longitude in present geodetic reference system at the beginning of the period.

Courtillot and Besse found a **magnetic pole reversal** correlation between the rate of axis movement and frequency of magnetic pole reversals. A fast episode of axis movement occurred 170-200 million years ago when the rate of travel was about 5 centimeters per year (assuming relatively steady movement over the period without cataclysmic surges). During the same period, magnetic reversals occurred at the rate of about 4-5 per million years. The number of magnetic pole reversals gradually decreased to zero some 165 to 10 million years ago, while the poles remained relatively steady. When the pole drift resumed its 4 centimeters per year rate, the reversal rate increased to 5 per million years.

There is a crude correlation between the velocity of pole movement and the decaying of reversal rates if a lag response is considered (the True Pole Wander appears to precede the core changes that are suggested to be the cause, or at least a contributor, of pole reversals). Courtillot and Besse, along with other scientists, suggested the correlation is related to the coupling between the mantle and core. They reasoned that high subduction rates that send cold slabs deep into the mantle cause instability in the magnetic field. This results in periodic pole reversals—not unlike the downdrafts and updrafts associated with violent atmospheric weather. Donna Jurdy, working with Gordon at Northwestern University, also proposed that sinking cold dense ocean plates and hot spots contribute to the dynamic changes. Periods when many hot convection plumes are raising up from the core-mantle boundary (offset by sinking cold slabs of crust) provide faster heat transfer by convection currents. Rapid heat loss by flood basalts, like the Deccan Traps and the Columbia Plateau, would also affect the magma current patterns.

William Seger, of Texas A & M, and Ulrich Bleil of the University of Bremen, found a cessation of pole wonder at about 40 million years ago that is not reflected in other pole motion plots. Courtillot and Besse suggested that the pole direction change 40 million years ago may be attributed to the collision of India with Asia; this consumed over 10,000 kilometers of subducting crust. At the same time the Pacific Plate took a 45 degree left turn as indicated by the alignment of hot spot created islands and seamounts of the Emperor and Hawaiian chains.

Courtillot and Besse speculated that mass extinction events are related to the pole dynamics and are based on climate disruption created by exceptional volcanism associated with periods of increased hot plume activity. They pointed out that the two major extinction events, Permo-Triassic and Cretaceous-Tertiary, occurred during periods of major volcanism that followed exceptionally long periods of non-reversal. The Deccan Trap in India (a lava flow, similar to the Columbia Plateau of Northwestern United States, covering over 200,000 square miles in Southeast India) erupted at the Cretaceous-Tertiary boundary. Another massive flood basalt bed, the Siberian Traps eruption, coincides with the Permo-Triassic boundary.

The Kerr and Courtillot-Besse articles mentioned above provide exciting food for thought when examined in light of the Dynamic Axis Theory. Even though some elements require more detailed analysis, few fundamental conflicts exist. The present True Pole Wander determined by Courtillot has the pole moving toward the 60 degrees east meridian. As noted earlier, present geodetic measurements indicate the pole is moving south along the 70 degree west meridian. This apparent discrepancy could be explained, if:

1. The pole makes another U-turn, so recently that it has not been detected in the

paleomagnetic data. For example, an event that triggered the termination of the last ice age a short 18,000 years ago.

2. Courtillot's True Pole Wander was distorted by the assumption that the hot spots are fixed. The existence of fixed hot spots would also alter the entire configuration of the 200 million year plot of the North Pole movement.

3. Not recognizing and eliminating from their computations the movements of the magnetic pole relative to the axis of rotation could introduce sizable errors in the data.

4. Ignoring and eliminating from their computations the local magnetic anomalies that affected paleomagnetic data.

It is exciting to speculate on the geologic story generated by axis drifts of the magnitude, as suggested by these noted scientific studies. It is also mind-boggling to speculate on the worldwide cataclysmic events associated with periodic surges, so let us take one step at a time.

Earlier studies

The speculation that the earth's axis moves is not new. The earliest attempt to define past pole positions were based on limited data. They failed to consider tectonic plate movements and other factors that adversely affected their accuracy. Before paleomagnetic data implicated the magnetic pole at various periods of geologic time, pole positions were selected based on their ability to explain evidence of ice caps and coal beds. This included placing the pole in India.

The introduction of paleomagnetic signatures that point toward the magnetic pole at the time the rocks formed triggered a new effort to define past pole wandering. The problem was to find rocks of a common age distributed on the globe in such a pattern, that the intersection of their pointing fingers provide a geographic position for the magnetic pole. Differential plate movements that distort intersection determinations were not taken into account. Any failure to detect and compensate for plate rotation would have been extremely damaging to paleomagnetic data for establishing pole positions. When we review the earlier studies, it is important to keep in mind their perspective and not to categorically reject everything just because their indicated pole movements were greater and even more erratic than those presently accepted.

Hugh Auchincloss Brown, an electrical engineer, was fascinated by the stories from the USSR about the find of a frozen woolly mammoth that was preserved over thousands of years. His search for an answer took many turns including the development of the HAB Theory (named after Hugh A. Brown) regarding sudden pole movements. Brown published his thoughts in the book "Cataclysm of the Earth". As an electrical engineer, Brown looked at the magnetic field as a key player in the game. While he recognized the tremendous energy of the magnetic field, it appears now that he miscalculated the physical mechanics of pole shifts.

It is not worthwhile to reconstruct Brown's deductions in detail. It should suffice to say that Brown believed that the Antarctica ice cap posed the threat. He felt that if the polar cap continued to build, it would make a rapid thrust toward the equator causing the pole to shift 80 degrees. His calculations were apparently based on a truly spherical planet, as opposed to the oblate ellipsoids. For the eccentricity of the polar ice cap to overcome the equatorial bulge and shift by 80 degrees, it must also stretch and compress the crust as discussed earlier. He did visualize that the pole shift would be cataclysmic to the physical planet, but not in terms of changes in the geoid or magma seas we have discussed. From chapters 2 and 3, we can see that even small shifts of only 1-2 degree will have far reaching consequences. The primary contribution to the Dynamic Axis Theory by Brown's book is his collection of geologic phenomena indicating cataclysmic events.

Charles H. Hapgood, a history teacher, delighted in challenging his students to do fundamental historic research. Student had to find out everything they could about the lost continent of Mu (the Pacific equivalent to the legend of Atlantis in the Atlantic). This launched Hapgood's interest in a study and evolved into his theory of pole shifting. The class project actually switched to Atlantis when insufficient information was found about Mu to serve Hapgood's class objective.

An article about Hugh A. Brown's theory in Argosy prompted Hapgood to study axis shifts to account for Atlantis' drowning by changing sea levels. Hapgood met with Brown, who shared his data freely. Hapgood, however, found he could not resolve some fundamental differences in dynamics, and consequently, wound up attacking the problem from a slightly different perspective. Hapgood published two books spelling out his concepts of polar movements: "The Earth's Shifting Crust", and "The Path of the Poles". In his first book he considered the ice caps as the trigger for the shift. Hapgood suggested that the pole shifted over an extended period and with a dramatically scaled down polar shift magnitude. Hapgood asked Albert Einstein to write the foreword to his book. Einstein was impressed by Hapgood's thoroughness and presentation.

Hapgood's second book responded to questions raised by Albert Einstein and Kirtley F. Mather, a geology professor at Harvard. Hapgood's second explanation for the pole shifts is based on the lithosphere slipping over the "wave-guide layer." The wave-guide layer is a very liquid layer of lighter rock material that was discovered by V. V. Beloussov, a Soviet geophysicist. The liquid layer, Hapgood reasoned, provided less shear resistance for the lithosphere to slip over the lower asthenosphere. He also looked at the wave-guide layer as the creator of gravitational instability that served as the driving force for crustal slippage. Hapgood's time scale for axis movements remained much slower than Brown's, as it was measured in thousands of years as opposed to a few days.

It is noteworthy, that for the lithosphere to rotate around the inner asthenosphere, with the only resistance provided by the wave-guide layer viscosity, it would need to be spherical in shape. Otherwise, all the problems indicated by the crustal tailor scenario in chapter 5 would still apply.

Hapgood's four pole positions were:

12,000 years ago	(90°N, ___)	present pole
18,000 years ago	(60°N, 83°W)	Hudson Bay (Wisconsin glaciation)
50,000 years ago	(73°N, 10°E)	Greenland Sea
80,000 years ago	(_____)	Yukon District of Canada

The above dates are presented only to show that Hapgood's data point to sizable pole movements in the past. The Dynamic Axis Theory is not in a position now to endorse or reject any specific pole path.

When I first read "The Earth's Shifting Crust", I felt I had finally found someone that had visualized the Dynamic Axis Theory. Each of his examples of geologic change and climate swings could not have set the stage better for the conclusions I am trying to present here. Rather than trying to paraphrase Hapgood's excellent work, anyone interested in further indications of axis movements should read his book from the perspective of the Dynamic Axis Theory.

Before we leave Hapgood, another one of his class projects is worth mentioning. Captain Arlington H. Mallory (an engineer, navigator, and author) planted the seeds for a study of ancient maps containing information that, according to recorded history, could not have been available to the authors of the maps. Hapgood concluded that a very advanced civilization existed long before recorded history. This could possibly relate to earlier civilizations responsible for many myths and legends. This fascinating research is published in his book "Maps of the Ancient Sea Kings: Evidence of Advanced Civilization in the Ice Age". Of special interest to our subject is the "Oronteus Finaeus Map" showing the outline of Antarctica, as if much of today's coastal ice was absent. The map was drawn in 1531, but the date of the surveys that furnished the data for the map construction had to precede that date. Hapgood speculated that the source data must have been handed down from much earlier times to account for its accurate portrayal of the Antarctica land mass at a time when considerably less ice existed.

Theoretical 1-Degree Axis Shift to (89°N, 70°W)

Records of polar motion show that the North Pole has been drifting south in the direction of 60-75 degrees west longitude toward New Brunswick, Canada. At the present drift rate of about 0.0032 seconds per year, the North Pole will shift:

     1-second in 312.5 years (~ 100 feet)
     1-minute in 18,750 years (~ 6,000 feet)
     1-degree in 1,250,000 years (~ 68 miles)

With no pole shift surges, this would place the pole in the vicinity of 89 degrees north latitude and 70 degrees west longitude (89°N, 70°W) in about 1.25 million years. Applying the notion of geoid change discussed in chapter 2 and the shift meridian values in Table B 1 (Appendix B, that assumed a rigid crust), we can calculate the approximate elevation change for any point on earth for this 1-degree hypothetical axis shift. Table B 5 (Appendix B) has a sampling of 186 elevation changes, many for coastal locations, to define the exchange between land and ocean bodies resulting from our hypothetical axis shift. The approximate values of geoid changes are derived by using the shift meridian geoid change for the latitude of each point. The shift meridian values are then reduced by using a parabolic prorate for the ratio of longitude distance of the point from the shift meridian to the neutral meridional great circle distance in degrees. Table B 4 lists the multiplier (percentage) to be used for each degree of longitude away from the shift meridian. This approximation is all that is justified in illustrating geoid changes, since the unpredictable crustal gymnastics would dramatically alter many of the real world surface elevations. Table B 5, therefore, only demonstrates the magnitude of sea level change that could be expected from a one-degree axis shift while assuming a rigid crust.

If crustal adjustments caused by magma pressure changes occur locally (near crustal weaknesses), the values computed for a rigid crust (Table B 5), could be close to the actual elevation changes for most areas of the planet. If, on the other hand, the crustal plates respond according to the notion of isostatic balance, the changes in sea level would be near zero. Indications of large and extensive elevation changes, as in the science fiction scenario discussed in chapter 1, tend to favor the first option.

Land-ocean configuration

The land-ocean configuration is dramatically altered by our hypothetical one-degree axis shift that moved the North Pole to 89°N, and 70°W. The shift meridians follow 70°W for the Western Hemisphere, and 110°E for the Eastern Hemisphere. The entire Asian continent falls in a positive quadrant where the higher geoid floods much of the lowlands around the edge of the continent. Asia's maximum change in elevation occurs at 45°N, and 110°E in Mongolia, where the higher geoid results in a lower elevation by 1,238 feet (377.3 meters). The other positive quadrant is centered at 45°S and 70°W in Southern Argentina.

The negative quadrants are centered at 45°N and 70°W in Maine, and 45°S and 110°E in the Indian Ocean southwest of Australia. The elevation change decreases in magnitude from the centers of the positive and negative quadrants to zero along the neutral meridian planes of 160°W and 20°E, plus the neutral equatorial plane and the poles.

A Closer Look at Sea Level Changes

It is possible to get a better feel for the affect of an axis shift on sea levels by taking a closer look at the flooding or drainage of each continent. Remember, however, that we are only considering the geoid changes.

Draining the North American continental shelves

For our hypothetical axis shift to (89°N, 70°W), all of North America, except Alaska west of 160 degrees west longitude falls within a negative quadrant. The lowered geoid exposes most of the present continental shelves. The alignment of the new Pacific Coast remains quite similar to the present with a few exceptions, because of the relatively narrow continental shelf along the Pacific Coast, and since it is several degrees west of the shift meridian. Canada's Inland Passage, Puget Sound, and San Francisco Bay are drained. Lower **California** is doubled in width as it picks up some continental shelf lands from the Pacific Ocean and some lands from the Gulf of California. Central America becomes much wider by exposing continental shelves on both its east and west coast.

The new Honduras and Nicaragua east coasts of Central America extend to the east for 100-200 miles. The Yucatan Peninsula extends about two degrees further north. The continental shelf off the east Texas and south Louisiana coasts are drained to move the coast line out about two degrees into the Gulf of Mexico. Florida picks up more than its present width from the Gulf of Mexico, while extending the state south to below the Florida Keys. The east coast of the United States from Florida to New Jersey gains about one degree of continental shelf lands from the Atlantic. An area about the size of

Pennsylvania is added off Cape Cod as Georges Banks is drained. Even more land appends to Nova Scotia by exposing the Sable Island Bank. Over 300 miles of new continent are added to the east of Newfoundland as the lowered geoid exposes the Grand Banks.

To the north, most of Hudson Bay is drained. Many of the islands of northern Canada attach themselves to the continent by new land bridges. Only western Alaska experiences flooding under the influence of the higher geoid of the Asian positive quadrant. An island about the same size as Florida forms by the drainage of the Great Bahamas Bank. All West Indies islands grow in size. Greenland gains land as its narrow continental shelves are drained.

Tilting of the Great Lakes Region

The elevation changes around the Great Lakes illustrate the consequences of an axis shift within a continental land mass (see Fig. 4.11). The 70th meridian, being the shift meridian, represents the high point for east-west elevation changes. The geoid drainage leaves the Saint Lawrence River 1,234 feet above sea level where it crosses the 70th meridian just east of Quebec, Canada. The river at Quebec is elevated by 1,237 feet (slightly more than at the 70 degree latitude crossing, because it is closer to the 45 degree latitude). The east end of Lake Ontario, near Kingston, Ontario, is raised 1,236-1,482 feet. The west end of Lake Ontario, at Hamilton, Canada, is raised 1,221-1,467 feet above sea level. This means that a Lake Ontario has to be 15 feet deeper at the west end before it flows out over the new higher east end.

The east end of Lake Erie at Buffalo, New York has an elevation of 1,792 feet (571' present elevation plus 1,221' from the lowered geoid). Lake Erie waters at Toledo must be 18 feet deeper to flow out over Niagara Falls. Lake Michigan, at Chicago, has to be 29 feet

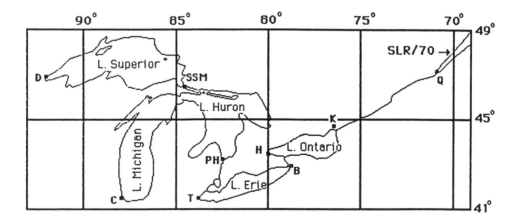

Fig. 4.11 Effect of hypothetical 1-degree axis shift on the Great Lakes Region.

deeper to match the elevation of Lake Huron. The 29 foot rise in Lake Michigan at Chicago will reopen the drainage out the Chicago Sanitary and Ship Canal to drain down the Des Plaines, Illinois, and Mississippi Rivers to the Gulf of Mexico. This is the same drainage pattern suggested for Lake Michigan during the Ice Age, when the St. Lawrence was still blocked by glacier ice.

Duluth, now the highest point, has an elevation of 1,763 feet (600' + 1,163'). This is 50 feet lower than the east end of the new Lake Superior at Sault Ste Marie. It will take a more detailed analysis of the surrounding lands to determine the lake boundaries and drainage pattern of the Great Lakes under the condition created by our hypothetical axis shift.

The elevated continental shelves around North America are vulnerable to erosion by the rivers that flow across them from the high lands. Deep, sharp canyons are soon carved in the newly exposed coastal plains. The lower Mississippi River cuts a deep sharp canyon through the relatively soft sediments elevated by just over 1,000 feet near the present delta, especially since part of the Great Lakes drainage joins the Mississippi drainage. In time, continued erosion creates the more rounded mature hills and valleys along the coast.

An elevated Europe

All islands in the North Atlantic become larger as the geoid drops in elevation. The geoid drop for Great Britain is much less than for eastern North America, but still enough for Ireland and England to join the continent. The balance of western Europe gains some new lands west of the 20 degree east meridian from the Baltic and Mediterranean Seas. Similar increases in land elevation will occur along the narrow continental shelves of Africa north of the equator and west of the 20 degree east meridian.

Flooding in Asia

Eastern Europe and all of Asia are flooded because of the higher geoid of the positive quadrant. With the elevation of Istanbul lowered by 232 feet, the Mediterranean Sea waters flow into the new deeper and larger Black Sea. A more detailed area analysis is required to be sure, but the enlarged Black Sea will probably join up with the even lower Caspian Sea to extend the new Mediterranean Sea deep into the Asian continent, possibly including the Aral Sea. We are told the inverse to this scenario occurred about 6 million years ago when the Mediterranean Sea dried up. Kenneth J. Hsü's article "When the Black Sea was Drained" offers more detail. An even more detailed account of Kenneth J. Hsü's research is presented in his book "The Mediterranean was a Desert: A Voyage of the Glomar Challenger". Chapter 8 includes some additional comments on his research findings.

The maximum flooding occurs along the 110 degree east meridian with inundation by sea waters flooding the lower Indus, Ganges, Mekong, Yangtze, Yellow, and Amur River valleys of southern and eastern Asia. Lowlands of the northern Soviet Union are also flooded, with an extended arm of the Laptev and Kara Seas reaching up the Lena, Yenisey, and Ob river valleys. The North Pacific islands west of 160 degrees west longitude loose lands or are completely inundated by the higher geoid. The Bering Strait is widened by the higher geoid allowing some of the warmer Pacific Ocean currents to circulate into the

Arctic Ocean to dramatically alter weather patterns downwind from the Bering, Chukchi, and Beaufort Seas. An axis movement in the opposite direction of our illustration would account for the land bridge forming between Asia and North America as assumed in the theory of migration to the new world.

Inundated coastal lands of South America and western Africa

In the Southern Hemisphere, a sliver of land along the west coast of Africa, west of 20 degrees east longitude and south of the equator, experienced some flooding. The South Atlantic islands would be smaller or submerged. South America, and the South Pacific east of 120 degrees west longitude all experience higher seas. Extensive flooding occurs in southern Chili, southern Argentina, Uruguay, Paraguay, and southeastern Brazil. Islands off the west coast of South America would be smaller, or completely submerged under slight flooding. Sea level of the Amazon Delta, being at the equator, would remain essentially unchanged.

Growth of Australia and eastern Africa

The negative quadrant of the Southern Hemisphere results in higher elevations for eastern Africa, Madagascar, and the islands of the Indian Ocean, including a large new island that emerges at the Seychelles Mauritius Plateau northeast of Madagascar. New Guinea and Tasmania is integrated into an enlarged Australia with land bridge ties. New Zealand's North and South Islands join, as its total land area doubles in size.

Shift in Antarctica's centroid

Wilkes Land in Antarctic is elevated and extends the continent to the north towards Australia by draining the narrow continental shelf. On the opposite side of Antarctica, some of the Palmer Peninsula and other lands bordering on the Weddell Sea are flooded. Antarctica remains ice locked. The loss of lands and glacier cap to the Weddell Sea, coupled with the extended ice cap over newly exposed Wilkes Land result in a shift in the centroid of the continent and ice cap. A decay of the ice cap on one side of Antarctica coupled with an expanded ice cap on the opposite side creates a weight shift that alters the earth's spin balance.

Other hypothetical axis shifts

To get a feel for the geoid change for any small axis shift along the 70-degree west meridian, simply multiply the values of [dG], found in Table B-5 (Appendix B), by the magnitude of the shift in degrees. The elevation of Denver after our 1-degree axis shift is 5,280' plus 1,033' = 6,313'. If we increase the shift to 5-degrees, Denver's elevation would be at 5,280' plus (5 times 1,033) = 10,345'. Table B-5 can also illustrate a 1-degree axis shift to (89°N, 110°E) simply by reversing all the signs. This opposite shift places the North Pole at (89°N, 110°E), and floods New York City under 6,110 feet of ocean waters, while Vladivostok is elevated to 5,795 feet above sea level.

Of course, all comments regarding flooding and drainage are based on the land areas and continental shelves as we find them today. To apply any axis shift to the past would require additional considerations for erosion and deposition that also dramatically alter the land-ocean configurations. Plate tectonic movements add another variable.

This chapter has addressed the effect of an axis shift on the geoid. It also provided us a feel for the magnitude of flooding and drainage for the geoid relative to today's crust. We know, however, that some of the elevation changes given in our hypothetical 1-degree axis shift will be offset by crustal gymnastics, as the earth's shell responds to the changing magma seas. With a better understanding of the effect, we are now ready to tackle the cause in chapter 5.

5   WEIGHT SHIFTS AND AXIS TORQUES

Each ton of soil, water, and air on the spinning planet creates a centrifugal force that establishes the earth's axis of rotation. The strength of the centrifugal force for each weight is a function of its distance out from the axis of rotation (spin radius). The centrifugal force acts perpendicular out from the earth's axis, like the water in a bucket that does not spill while being swung over your head. Gravity provides the centripetal counter force that holds objects to the planet, just as your arm keeps the swinging bucket from sailing off like a rock released from a slingshot. Gravity forces also facilitate the stream erosion process by pulling the sediment laden streams toward their deltas. For a river, such as the Mississippi, the eroded sediments move toward the equator and are deposited further from the earth's axis to increase the centrifugal force. The reverse is true for rivers, such as the Ob in the old Soviet Union where sediments are deposited closer to the North Pole. Our planet responds to each movement of weights on or within the planet by altering the angular velocity of the spinning orb or disrupting its spin balance.

Dynamic balance

Every element added to or removed from our spinning planet, except those on the equator and the spin axis, create a force couple that tries to realign the unrestrained earth's rotating axis. The distance from the equatorial plane measured parallel to the earth's axis, provides the moment arm (lever arm) for the force couple. Consider a loaded truck parked on the equator. Being as far from the axis as it can get, the centrifugal force on the truck is at a maximum. Its moment arm is zero, consequently, it does not produce a force couple on the earth's axis. If we move the truck to one of the poles, its moment arm is at a maximum, but its centrifugal force is reduced to zero since it is on the axis. The maximum force couple, created by our loaded truck, will occur when it is at 45 degrees latitude where both the centrifugal force and moment arm are at about 70 percent of their maximum values. A more detailed discussion of the centrifugal forces and force couples that act on the earth's unrestrained axis is included in Appendix C.

The growth and decay of continental glaciers can produce large centrifugal force couples if they are eccentric to the poles. This contradicts Brown's theory, that suggests that the buildup of the Antarctica ice sheet would trigger the tumbling of the earth—or is it? We normally think of Antarctica as being centered over the South Pole. If it was, force couples created by growth or decay of the ice sheet would be counterbalanced. However, a decay on one side of Antarctica in concert with a growth on the opposite side, as suggested by our hypothetical axis shift example, will create an additive force couple of considerable magnitude. The primary factor limiting the expansion of Antarctica's continental glacier is its island type platform. Nevertheless, any growth of the ice sheet on one side of Antarctica combined with an equal decay on the opposite side will create a resultant force couple twice as large as created by equivalent growth or decay of Greenland's ice sheet. Greenland is obviously eccentric to the earth's axis and its centroid is about the same distance from the North Pole as the outer edge of Antarctica is from the South Pole. Increasing the thickness of the ice sheet in southern Greenland would create the greatest torque.

The ice sheet that covered much of North America during the last Ice Age compares favorably in area and thickness to the present Antarctica ice sheet. Its centroid, in terms of today's spheroid, was about twice as far from the North Pole as Greenland's centroid. The force couple produced by the North American ice sheet had some help from the western Europe ice sheet. Since no equivalent counter force couple existed in Asia or the Southern Hemisphere, nearly the full torque of the ice sheet force couple would have been applied to the earth's axis. Here we need the services of an expert in dynamics to calculate the response of our spinning orb. To make the computation assuming a rigid crust is relatively simple. To make the computation that includes estimates of the magma flow and crustal reaction to an axis shift is far more speculative and challenging.

Isostatic and isothermal-pressure balance

This is as good a time as any to detour slightly and discuss the concepts of isostasy or isostatic balance and a companion isothermal pressure balance. Isostatic balance is based on the well understood Archimedes' Law. Simply stated, when an object, such as a boat floats in water, the weight of the volume of water displaced is equal to the weight of the boat. A loaded boat simply floats deeper in the water, it displaces more water.

According to the isostatic balance theory, the added weight of the ice sheet that grew over Canada pushed the crust deeper into the mantle. When the glacier melted, the crust rebounded. Recent changes in sea level along the northeastern coast of the United States and the Hudson Bay Region are still attributed to rebounding long after the ice sheet has gone. However, where did the displaced magma go?

If we accept isostatic rebound from unloading the ice sheet as the cause for today's sea level gauging stations being elevated, it shows that a considerable lag time does occur. The same lag in time would have occurred as the crust subsided during the ice sheet growth. Similar time lags are associated with the subsidence of the crust responding to the sediment deposition at the river deltas and a corresponding rebound of the eroded continent. Before the construction of the Aswan Dam on the Nile, the deposited sediments continued to build up the Nile Delta, even while the area was subsiding because of the added weight. With the sediment source cut off by the Aswan Dam, an invasion by the sea is now threatening the Nile Delta as it continues to subside. The lag in the crust's response to changes in crustal loading creates a period of dynamic imbalance. The imbalance induced by an eccentric weight will continue to act until isostatic balance is restored by the earth's seeking a new axis of rotation.

The lithosphere floating on the magma seas does not respond exactly the same as a boat floating on the ocean waters. To start with, the crust is continuous over the entire magma sea, and it has enough internal strength to distribute a spot weight over a much larger area. When one crustal area is pushed deeper into the magma interior, another area must be raised by an equivalent amount. This is similar to the response when we punch a bean bag or lay on a waterbed.

Lets limit our scope of consideration, for the moment, to continental depression under the ice sheet and a corresponding rebound elsewhere on the planet. If we assume the added ice sheet weight, deposited as snow, came from the ocean basins, it follows that unloading of ocean crust will cause the sea floor to rebound an equivalent amount by weight. It makes no difference if the ocean floor bows up over large areas or locally as blisters. In either case, sea level around the world will remain essentially unchanged. The only evidence of sea level change, consequently, would arise from the subsidence of the continent as it is driven deeper into the magma by the ice sheet. Everything else being equal, marine terraces formed in the ice sheet region would indicate higher, not lower, seas along the coast of northern North America and western Europe when more water is locked up in continental glaciers. One additional response, however, comes into play. When the ice weight is added over the central continent, the displaced magma will resist being pushed out uniformly under the ocean basins because of the high magma viscosity. This increases magma pressure locally around the edge of the continent and may cause it to rebound. Consequently, marine terraces formed during the Ice Age, could be either higher or lower ,depending on the crustal warping.

The Mohorovicic Discontinuity (MOHO), the surface separating the crust and mantle, has been explained by isostatic balance. It is true that the depth of the MOHO does reflect the lithosphere relief, being higher under the oceans and deeper under the continents (Fig. 5.1). We are told that the combined weight of the thinner, but heavier

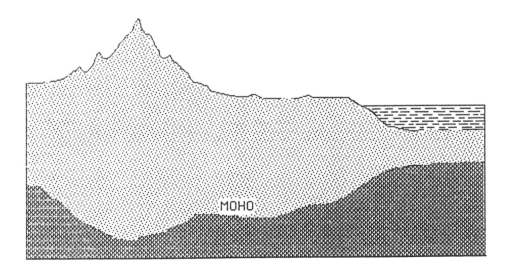

Fig. 5.1 Mohorovicic discontinuity (MOHO).

ocean crust (greater density rock) and ocean waters floats higher on the magma sea than the thicker, but lighter (less dense rock) continental crust. Where mountain ranges extend high above the rest of the continent, a companion root extends deeper into the magma sea.

The MOHO, however, represents the temperature-pressure surface at which the crust looses its brittle characteristics. The intense pressure and higher temperatures below the MOHO, allow the ductal upper mantle rocks to be remolded. This is comparable to a blacksmith when he forms new shapes in heated steel. Even deeper in the earth the temperature-pressure reaches the melting point of crustal rocks to form magma. Magma may be equated to a ladle of molten steel at a foundry. The magma pouring out of Kilauea Volcano provides us with a glimpse of the molten magma from the earth's interior. Pressure on the earth's interior serves the same role as a pressure cooker in that it raises the boiling point of the magma. The pressure is reduced as magma moves up through the vent of a volcano. The reduced pressure allows the magma to remain fluid even if it cools slightly. When the intruding magma looses enough heat to be cooled to below a critical temperature-pressure, it freezes—either within or on top of the earth's crust.

Vertical convection currents in the core and lower mantle transfer heat up from the earth's interior furnace. Hot magma currents rise, as cooler magma sinks into the magma seas to complete the vertical circulating eddies. This resembles the vertical atmospheric eddies that form cumulus clouds from summer's heat. When the hot currents of the magma seas contact the pliable upper mantle, or rigid crust, most of the heat transfer switches from convection to conduction. The earth's internal heat continues its travel outward as a function of the temperature differential and the coefficient of heat transfer of the crustal materials. Heat transfer switches back to convection through the ocean waters, and the atmosphere. The planet eventually looses heat to space by radiation. Of course, radiation acts in two directions, as heat radiated by the sun also warms our planet.

Consider what would happen if we could cover the United States with a perfect insulation to trap all the heat that escapes from the magma seas. Eventually the trapped heat would melt the crust up to the insulation cover—just as the winter ice on a lake would melt if it were covered by an insulating cover.

Now consider what would happen if the thinner ocean crust became exposed to the atmosphere by an axis shift. The ocean crust, exposed to the atmosphere, would thicken as the increased heat loss promotes crustal freezing—like ice forms on a lake when heat is lost to the colder winter air by convection and radiation. Conversely, the thicker continental crust, converted to a deep ocean basin, would partially melt from below and become thinner. This process can explain the exchange of continental lands and ocean basins.

Although isostatic balance does play a role in defining the MOHO, it is not the controlling factor as suggested by many textbooks.

Cause of the Axis Wobble and Axis Drift

Our planet would spin true on its axis if it were in perfect balance. The axis wobble cycles would not exist if a force couple of some kind had not acted on the earth's axis. If a force is introduced to cause the axis to wobble, then ceases to act, the physical properties of dynamics predict that the wobble will dampen out in time. Considering the limited period for which pole motion records have been recorded, signs of dampening have not been detected. In searching for force couples created by weight shifts that would cause the Chandler and annual wobbles, it soon becomes evident that weight shifts from a combination of many sources must be involved.

Manufactures of machinery with rotating parts go to great length to balance the system to prevent undesirable vibrations or wobble. Tires are a crude example with which most of us are familiar. For instance, knot can form on a tire, or it wears unevenly causing the imbalanced tire to thump down the road. Tire balancing is usually accomplished by adding, or shifting, lead weights attached to the wheel rim. One major difference between the rotating machine parts and the rotating earth is that machine parts are restrained by bearings or casings and an axle, while the earth's axis has complete freedom to wobble and seek a new axis of rotation.

It is not difficult to identify many natural forces acting on the earth. Our planet's natural computer (enforcing the laws of Nature), keeps track of all the weight shifts and reacts accordingly, even if the reaction is too small to be measured by our most precise and sensitive instruments. Only a few of the forces, other than gravitational pull of the sun, moon, and other planets come from outer space. Small meteorites and meteoric dust are attracted to our planet as it works like a vacuum cleaner sucking in space debris that intercepts its path while orbiting the sun. Attracting only a few small meteorites may not have always been the case. The sun's solar system may periodically pass through regions filled with larger galactic debris that bombards our planet. The moon's surface, as well as images returned by our probes of neighboring planets, supports such a hypothesis. Louis Alvarez and his son, Walter, proposed that a large asteroid or volley of asteroids struck the earth and caused periods of mass extinction—including the demise of the dinosaurs. We do not, however, have to look beyond the dynamically active earth to find many cases of weight shifts that affect the spin balance and alter the axis of rotation.

Kinetic energy of earth's rotation

The earth traveling in its orbit around the sun represents a tremendous amount of kinetic energy, like a runaway freight train. This energy remains untapped until it meets resistive forces. The planet's rotation on its axis also represents a tremendous amount of kinetic energy. When the axis shifts, it changes the centrifugal forces acting on the ocean waters and magma seas, which in turn alters land-ocean boundaries and distorts the crust. The kinetic energy of the orb's spin is, therefore, tapped to accomplish work, such as driving tectonic plates, earthquakes, volcanic eruptions, altering the erosion patterns of land masses, disrupting atmospheric currents, and changing the flow patterns of the oceans.

Weight Shifts and Wobble Cycles

If we accept the basic notion that a free spinning orb will respond to changes in the dynamic balance, we need to identify contributing natural weight shifts. For convenience, let us consider weight shifts for the biosphere, atmosphere, hydrosphere, lithosphere, and asthenosphere. The identified weight shifts will very from insignificant to substantial. The exercise, however, brings into focus the variability in the earth's dynamic balance.

As we identify shifting weights, it is important to catalog each by their cyclic influence.

It seems reasonable to assume that the annual wobble cycle is influenced by seasonal activities, but the 14.29-month cycle does not fit a similar natural cycle. After identifying the players in the game, we should be in a better position to speculate on the cause of a 14.29-month wobble cycle. Accumulative weight shifts are also important clues to axis drift and the probability of axis surges.

Biosphere weight shifts

We can search for natural forces acting on our planet by analyzing some of the weights that shift in the biosphere where humans are walking, running, driving, and flying in all directions. On any given day, such as the Super Bowl or the Olympic Games, many people concentrate near one spot on the globe. Humans are joined in this undirected marathon by all other species of animal life in the world, including the migratory herds of caribou and flocks of geese. Human activities include the extraction of oil and gas from wells, minerals from mines, and trees from forests. The product of our efforts often concentrates the weights where we build cities and reservoirs.

Among the weight shifts noted above for the biosphere, the north-south animal migration provides a seasonal weight shift. Land-ocean distribution insures that movements on complementary longitudes (180 degrees apart) will seldom provide offsetting force couples.

Plant and animal relocations due to major climate swings create a one shot weight shift that may last for years. Even then, since the materials for plant and animal life are extracted from the soils, little or no effective weight shift is involved. Human engineering activities of building reservoirs and cities are additive and contribute to some non-annual cycle and the axis drift.

The magnitude of the biosphere force couples, however, is so minute that by themselves they must create only small blips in the earth's axis wobble. Although biosphere activities are a part of the weight shifting equation, their influence on the earth's spin balance is dwarfed by other considerations discussed below.

Atmosphere weight shifts

Kelvin, an Englishman, directed his thoughts towards physics while America was engaged in the Civil War. He visualized that as the atmosphere high and low pressure areas (density variations) move around the globe, they represent a change in the physical loading on the planet that affects the earth's rotational balance. The daily movements of a high pressure cell in the atmosphere can be thought of in the same terms as the hypothetical shifting of Mount Everest used in the Appendix C illustrations. We may not think of air as having weight since we are so used to having it around. The change in atmospheric pressure may not even affect us much when we move from sea level to higher elevations. A half full plastic bottle of hair dressing, securely capped in Denver, reminds us of the atmospheric pressure difference when we unpack it at sea level and find the sides of the bottle have caved in. On the return trip, the pressure released when the cap is removed at a higher elevation provides a similar reminder.

High pressure atmospheric cells over ocean waters, like a loaded boat, displace ocean waters (locally lowering the sea level). Water is pushed away from the high pressure cells to create higher sea levels under the low pressure cells. High tides associated with the very low pressure of hurricanes are direct evidence of sea level's response to variations in atmospheric pressure. The relatively low viscosity of ocean waters makes their response time for isostatic recovery to atmospheric loading relatively quick. On the other hand, when a high pressure atmospheric cell moves over land areas, the added weight increases the load on the lithosphere boat floating on the magma seas. The full isostatic response of the high viscosity magma seas may never be achieved before the cell moves out of the region. Jerome Namias, a meteorologist at Scripps Institute of Oceanography, advanced a theory involving the effect of air pressure variations on the San Andreas earthquakes (discussed in chapter 8).

The atmosphere high and low pressure areas move so rapidly across the globe that they influence the daily and weekly magnitudes of axis wobble. Rapid movement of atmospheric highs and lows are probably responsible for much of the noise (erratic variation) observed in the raw polar motion data. Except for the precipitation transported by the atmosphere (discussed under the hydrosphere paragraphs), no accumulating force couples are generated to drive an axis drift.

Seasonal temperature flip flops between the north and south hemispheres would have little or no effect on the axis if they were uniformly distributed around the world at each latitude. Because of the land-ocean distribution, however, the seasonal heating and cooling of atmospheric current are not mirror images between hemispheres. Consequently, seasonal atmospheric currents can provide some annual influence on the earth's dynamic balance. One case in point is the thermal low that sets up over the Mojave Desert during part of the summer months. It is replaced by a relatively stationary high pressure cell during much of the winter. These parked pressure cells provide a steering influence on the jet stream and weather by deflecting normal atmospheric circulation paths to the north in the winter, and to the south in the summer.

Friction between prevailing atmospheric winds and the earth's surface can affect the earth's spin rates. westerly winds buffeting the Rockies accelerate the earth's rotation. Winds blowing across the Southern Rockies have a greater effect on the spin rate than the same winds blowing across the Canadian Rockies, simply because of the greater lever arm (spin radius).

One climate factor that exceeds a one year cycle is global or hemisphere cooling and warming, possibly determined by outside sources. The Milankovitch cycle hypothesis (discussed in chapter 8) fits into this possibility. A series of large volcanic eruptions or asteroid strikes that cast debris into the upper atmosphere and stratosphere to blot out the sun's heat have been advanced as major contributors to cooling world climates. Although a series of large volcanoes would dramatically alter the climate for a few years, it is more difficult to visualize a geologic age—spanning thousands or millions of years—being fueled by a continuing series of large volcanic eruptions.

The interplay between the different spheres prevents any clear-cut assessment of their influences. Prevailing atmospheric current patterns are altered by major realignments of land-ocean areas. The atmosphere teams up with the hydrosphere to complete the hydrologic cycle. The weight shifts in the atmosphere, like the biosphere, appear to play only a minor role in altering the earth's axis.

Hydrosphere weight shifts

Water in its many forms makes up a substantial part of the shifting weights on our planet. Not all, however, significantly alter the force couples acting on the earth's axis.

The hydrologic cycle evaporates moisture from both oceans and land surface areas. Moisture laden atmospheric currents deposit their load some other place on the planet as rain or snow. Rain waters usually make a hasty retreat back to the oceans, but not before contributing to the soil moisture and aquifers. Snow stays around a little longer before making its way downhill. Variations in inland lake levels alter their spot weight. The present Great Salt Lake provides just a fraction of the weight that existed when Lake Bonneville was full (1,000 feet above the present lake level, and covering about 20,000 square miles). The same applies to Death Valley and other depressions that were lakes during wetter periods. Artificial reservoirs represent spot weights added during modern time.

Ocean currents, like atmospheric currents, contain high and low density cells. Temperature and salinity variations contribute to the ocean water densities that account for weight shifts as cells circulate within and between ocean basins. Today's deep ocean waters are cold except where heated by upwelling magma. The heavier cold waters from the polar region sink down as part of the vertical circulation patterns within the oceans. The Mediterranean Sea provides a second vertical circulating trigger. Because of its higher concentration of salts, the heavier water from the Mediterranean flows through the Strait of Gibraltar and dives into the depths of the Atlantic Basin. R. W. Stewart's "The Atmosphere and the Ocean" is one of many interesting articles on the subject.

We are more familiar with the horizontal circulating ocean patterns, such as the Gulf Stream that circulates warmed ocean waters up the western Atlantic and across toward Europe. Some of the warmer waters are peeled off to become part of the Arctic circulating eddies. The bulk of the Gulf Stream waters becomes the cooled current that flows south along the eastern Atlantic in its return to the tropics. Of increasing interest to oceanographers and climatologists are the relatively warm El Niño and cool La Niña circulation patterns of the Pacific.

Oscillating changes in sea level in the Pacific waters have been detected by tide gauges. However, many gaps exist in the tide gauge records simply because of the lack of islands to establish a vertical reference. The Unites States Navy's GEOSAT and National Aeronautics and Space Administration's SEASAT satellites with radar altimeters permit measurement of elevation changes in the open sea surfaces. The satellite radar altimeter data matched the tide gauge data within about one and one-half inches. They provide a viable alternative, especially since they offer more complete data throughout the open seas.

What did they find? First was the movement of a ridge of water driven by seasonal winds. Late each year, the band from 8 degrees north to 20 degrees north latitude raises as much as 4 inches at the expense of a band from 7 degrees south to 7 degrees north latitude. A reverse shift in sea level occurs just before midyear. No companion sea level cyclic change was found in the Southern Hemisphere. The above cycles were observed in 1985 and 1986.

In 1987 a different sea level shift, referred to as the Kelvin wave, was noted. The Kelvin wave is a 6-mile-wide wind driven wave about 8 inches high that moved from west to east across the Pacific. Scientists are trying to tie down any relationships between the east-west sloshing of waters across the Pacific, and the El Niño-La Niña cycles. Richard A. Kerr's, "The Big Picture of the Pacific's Undulations", provides more detail on this.

Glaciers are the massive accumulations of moisture, in the form of snow and ice, that change the weight balance as they grow and decay. While some valley glacier is receding at a rate measured in miles per decade, another glacier in the same general region may continue to grow. This apparent contradicting evidence in regional climates simply points out the role local changes in hydrosphere and atmosphere patterns play in determining the whole world climate.

The present ice sheets in Greenland and Antarctica are not permanent features. Some evidence that support both growth and decay of today's glaciers is constantly introduced into the debate. The huge ice sheet that covered most of Canada and northwestern Europe was at its peak only a short 18,000 years ago. One of the latest advances of ice occurred less than 12,000 years ago (recent changes in the carbon-14 dating techniques may push the dates a little further back into geologic history). Only a few thousand years later the ice sheet disappeared. The continental ice sheet melt down provided a large weight shift over a short geologic period. Even though the decay was driven by an annual cycle of summer melting interrupted by winter cooling, the long-range effect was additive over the meltdown period.

Ocean tides caused by the gravitational pull of the moon and sun create weight shifts with a cycle just over half a day in length. Since the moon's orbit around the earth and the earth's orbit around the sun are elliptical, the magnitude of gravitational pull is not constant. When both the moon and sun are on the same side of the earth, their gravitational pull is additive. When they are at right angles (quarter moons) the gravitational pull from the sun and moon provide independent gravitational attractions.

The prevailing northeast trade winds at the latitude of the Panama Canal Zone create a higher sea level on the Caribbean coast than the Pacific side. Much of the excess budget of ocean waters on the Caribbean side would be eliminated by flooding of the Central American land bridge between North and South America. Tsunami provide short term changes in the dynamic balance as they race out across the ocean from strong earthquakes. An increase in the earth's rotation rate adds to the equatorial bulge and polar deficit of the ocean waters. Since the oceans are not symmetrically arranged on the globe, any rotational changes affecting the geoid creates eccentric weights that act on the earth's axis.

The hydrosphere's prevailing ocean currents and eddies, like the atmosphere, have a seasonal variability within and between the hemispheres. The resulting weight shift in ocean currents has a greater seasonal bias than the atmosphere because of the restrictions imposed by the land-ocean configuration. As for the atmospheric currents, most of the present ocean current changes are under the influence of short term cycles. They do not appear to provide an additive long-term contribution of weight shifts to drive an axis drift.

Each change in land-ocean distribution, provided by the Dynamic Axis Theory, establishes a new prevailing ocean currents' pattern. The closing or expanding of the Bering Straight will affect the exchange of waters between the Pacific and Arctic Ocean and the moisture content of atmospheric currents passing over North America. Flooding of Panama and part of Central America would change the normal ocean current patterns in the central Pacific and Atlantic. Inundation of large continental land areas expands the continental shelf. Circulation patterns of shallow sea waters are considerably different from circulation over deep ocean waters. Draining of continental shelves likewise alters the ocean current pattern.

A major change in ocean current patterns provides a good example of the domino effect triggered by an axis shift. There is a direct relationship between ocean current patterns, evaporation rates, atmospheric circulation patterns, and annual downwind precipitation rates by region. Precipitation rates, in turn, determine inland lake levels, soil moisture, and glacier deposits. The long term effect could even be the growth or decay of a continental ice sheet.

To summarize: snow fields, glaciers, lakes, and soil moisture fluctuate on a seasonal basis to provide direct weight shifts that contribute to the annual wobble cycle. The long term accumulation or decay of glaciers affect some cycle other than annual. The meltdown of the ice sheet covering much of Canada and western Europe, over a period of about 6,000 years, constitutes a very sizable weight shift over a geologic minute. The greatest weight shift probably occurred during the early years of the meltdown, simply because more surface area was available for melting. Both the non-annual wobble cycle and axis drift are the product of additive long term weight shifts.

Erosion, as a product of the hydrologic cycle, is deferred to the discussion of the lithosphere that follows.

Lithosphere weight shifts

When we think of shifting weights on the planet, changes in the lithosphere usually stand out. Nature has provided us with an erosion process that attempts to reduce all land surfaces to sea level. Mud and rock slides tumble down the sides of alpine mountains. Streams and rivers carry tons of debris to their deltas in the ocean, inland lakes, or arid basins. Winds transport dust and sand particles that are usually deposited in relatively low areas where they are vulnerable to stream erosion. Some of the air borne dust, sand, and volcanic ash settle directly on the open seas. Volcanic eruptions, such as those of Mount St. Helens, Tambora, and Krakatau accelerate the local erosion process by literally blowing the tops off volcanic mountains and spreading volcanic ash downwind.

The world's land masses, however, are not in danger of extinction. Volcanoes must grow through a series of eruptions before they can be destroyed by an explosive eruption. Only relatively few volcanoes suffer the fate of Tambora. Lava flows and volcanic ash deposits buildup land masses. Mountain uplifts that are a result of magma intrusions or tectonic plate collisions make a significant contribution to renewing continents and islands.

The sedimentary beds that make up continental shelves are active evidence of the recycling of land masses. Tons of sediments are being added every second as land areas surrender sand and silt. Most of the thick sedimentary beds of the present continents, such as the Colorado Plateau, the Great Plains, and the folded Appalachian Mountains were laid down in shallow seas of ancient continental shelves. Changes in the geoid, as discussed in chapter 2, provide a logical explanation for how shallow seas can be interchanged with highlands. The hypothetical 1-degree axis shift demonstrates how the sediments now being deposited on continental shelves of North America, western Europe, and Australia, can easily be added to the continent as coastal plains. At the same time, the lowlands of Asia and South America become shallow seas that accumulate additional sedimentary beds. Alternating such regions between exposed lands and shallow seas account for the discontinuity in geologic periods as recorded in the sedimentary beds. The big question is, how do erosional weight shifts relate to the 12- and 14.29-month cycles?

Although spring flooding and summer cloudburst generated erosion has a seasonal cycle, the major effects of stream erosion are additive year in and year out. Stream erosion patterns, as discussed earlier, are cumulative as long as the land-ocean configuration and river drainage basins remain essentially unchanged. Since erosion is an additive event, it may account for the agitation forces necessary to maintain the axis wobble. The additive force couples of the South flowing rivers in North America and north flowing rivers of Asia can also account for at least part of the axis drift associated with the axis wobble (more detail is presented towards the end of this chapter). The yearly variation in erosion volume from river basins accounts for some of the erratic patterns and the noise of axis wobble records.

Volcanic eruptions may either create growth (add weight) or destroy volcanic cones (reduce weight). As they are isolated events that erupt on a random time scale, they would not provide a predictable pattern change in the wobble trace. Positive magma pressure that moves up through the crust to form a volcanic cone or flood lava can be attributed to magma tides flowing from the negative to positive quadrants. A slight increase in the centrifugal force comes from the elevated lands being at a greater radius from the axis of rotation, thus contributing to a change in the force couples.

The continuing growth of mountain ranges, e.g., the Himalayas, like stream erosion, would produce a more patterned, long term contribution spin imbalance. Other more subtle uplifts and subsidence, that involve large regional or continental areas, may provide equivalent or even larger force couples.

Asthenosphere weight shifts

We know less about the asthenosphere than the lithosphere, hydrosphere, and atmosphere simply because we cannot physically see it or directly measure its changes. We can, however, speculate on the makeup of the asthenosphere based on our understanding of natural laws as they apply to things we can see and measure. Modern technology is providing more remote sensing tools that allow us new insight into many of our planets interior mysteries. Our physical evidence includes the volcanic eruptions that pour out molten magma. As we drill deeper into the crust, the temperature increases. Extrapolation of the known temperature and pressure increases to deep within the earth indicates sufficient temperatures would exist to convert crustal rock into molten magma.

A slight digression allows us to raise the question of how much pressure exists at the earth's center. If we assume that each element within the earth is supporting all the elements above it, pressure must increase with depth. If we assume the mass that each element supports is the weight of an overlaying inverted cone, the pressure would be greater than if it supports only a cylinder of mass above the element. Gravitational pull (weight) produced by the earth's mass acting on a small ball of mass at the earth's center, however, is zero. Each element, from the earth's surface to its center, is subjected to gravitational pull by the earth's mass above and below. Consequently, the net gravitational attraction reduces with depth.

Earthquakes require some brittleness in the rock to build up stress. Since the focus of all earthquakes occurs within a few miles of the surface, with none originating deeper, this supports the magma core theory. The deepest earthquakes occur along subducting zones where blocks of the colder crust are being driven down into the magma seas.

Seismic waves tell scientists much about the inner layers of the planet since they react differently as they reflect, refract, or change velocity through each layer. Careful and controlled seismic studies during the past few years have revealed some of the physical properties deep within the earth. Scientists have identified critical surfaces within the planet where seismic waves abruptly change velocity, are reflected, or their direction is slightly altered (refracted). By analyzing multiple seismic wave velocities and reflections at several locations, scientists can obtain a three dimensional picture of the earth's interior. This resembles CAT scanners that are used to view inside the human body without physical intrusion. Using supercomputers, tomographers can convert the collected seismic data into 3-D images. One 3-D image application is for tomographic maps that show rising magma regions and cold sinking regions of magma, as presented in Jim Dawson's article "CAT Scanning the Earth" in the May 1993 issue of Earth.

Although there has been general agreement about the makeup of each concentric layer from the crust down to the inner core, the new data indicate that the surfaces between layers are not as smooth as once believed. In fact, relief features much larger than our mountain ranges and ocean basins, have been detected at the lower mantle-core boundary.

The biggest unknown in our shifting weight and force couple equations is the asthenosphere. Speculation about the existence of magma currents is the foundation of the most popular hypothesis used to explain the forces behind tectonic plate movements. The hypothesis, using primarily a vertical circulation pattern, calls for the upwelling of hot magma along the spreading plate edges. The upwelling magma is assumed to flow in both directions away from the spreading ridge, thus serving as a conveyor belt to transport the tectonic plates. The magma current looses heat through the crust and eventually becomes cooler and heavy enough to dive back into the lower mantle at subduction zones. Along the deep ocean trenches or at the junction between two plates in collision, the downward magma current carries with it part of the crust.

Evidence that contradicts the conveyor belt theory will be discussed near the end of this chapter.

Magma sea circulation patterns are open to speculation. The geophysicists view the upper mantle as a pliable solid—not brittle like the crust or fluid like the lower mantle. The increased heat and tremendous pressure modify the lower mantle's fluid attributes. The term "magma sea" as used in this book, refers primarily to the lower mantle, but recognizes the plasticity of the upper mantle as being a part of the crustal massaging process.

A molten magma asthenosphere should have many of the circulating characteristics of the atmosphere. Its high viscosity reduces the velocity of magma currents, but the patterns should be similar to the atmospheric currents. The nuclear ovens, assumed to be active deep within the earth, provide the necessary heat differential to drive the vertical and horizontal mixing of high and low density cells of magma currents. The hot spots (such as the one that has torched its way up to the surface and is assumed to be the source of a string of volcanoes that include the Hawaii Islands) just might be the asthenosphere's counterpart to the atmosphere's tornadoes. There extreme heat differentials create the tight spiraling magma currents that lift the hotter magma to the top of the asthenosphere.

The earth's rotation sets up the prevailing horizontal circulation pattern of the atmosphere from the equatorial doldrums to the polar cells. It includes the Coriolis effect on low pressure circulation, such as hurricanes and tornadoes (counterclockwise in the Northern Hemisphere, and clockwise in the Southern Hemisphere). The earth's rotation furnishes the same driving forces to the magma seas. The Coriolis force should establish the circulation patterns both for the core and mantle. This is consistent with the hypothesis of hot spots in the asthenosphere having a similarity with tornadoes in the atmosphere. Friction between the circulating magma and its upper mantle cover provides additional resistance to slow the fluid movements within the magma. A similar friction surface is not present on the upper surface of the atmosphere, where high velocity jet streams race around our planet. Even with these differences, the comparison is made with the atmosphere, as opposed to the hydrosphere. The circulation within the asthenosphere, as for the atmosphere, is not restricted by the continental barriers that interfere with ocean currents.

Although the high and low density (pressure) areas of the asthenosphere move at a snails pace compared to the atmosphere, the shift of mass involved is much greater. Gravity undulations in the geoid indicate the varying densities, not only of the crust, but also within the asthenosphere. Regardless of the current patterns, any variations in the magma cell densities circulating below the crust will create a weight shift. Not every cubic mile of magma has the same weight. When weights shift within the crustal shell, it is like a load shifting within a truckload of cattle that disrupts its dynamic stability. Considering the mass involved, the asthenosphere just might be the prime contribution to dynamic balance changes. There is no apparent seasonal connection with the asthenosphere.

A debate on how the mantle circulates heated up when 125 earth scientists gathered at California Institute of Technology in May of 1991. At issue was whether a Layered-mantle circulation pattern is confined to the upper mantle or if the Whole-mantle circulation pattern reaches the earth's core. Richard Kerr's article "Do Plums Stir Earth's Entire Mantle?" presented some of the debate. The Dynamic Axis Theory can accommodate either boundary, but favors the Whole-mantle circulation notion.

One of the objectives of examining shifting weights on our planet was to determine the contributors to the identified wobble cycles. Some cycles such as precipitation, temperature, and animal migration must contribute to the annual cycle of axis wobble, since they oscillate on a seasonal basis. A reason for the 14.29-month cycle in the earth's axis wobble is not as apparent. Since no direct 14.29-month oscillation of shifting weights has been found, it appears that the present 14.29-month cycle is determined by the earth's accumulation of eccentric weights (state of imbalance from non-annual cycles). In ages past or in the future, the 14.29-month wobble cycle period may be much longer or shorter.

Although 12- and 14.29-month periods are the two primary cycles that have been identified, other cycles should not be ruled out. A 42-year cycle, which appears to be a product of 12- and 14.29-month cycles, was noted earlier, as were minor contributions by the 2- and 9-week cycles. Additional cycles, presently assumed to be data noise, may be isolated when better data are available. Scientists normally proceed from the known to the unknown. Known weight shifts can be used to calculate their effect on the axis of rotation. A reverse computation, using the known axis wobble and drift can identify the unknown weight shifts that have an impact on our planet. One of the primary targets of this approach is to determine the influence of magma currents.

Since the axis wobble will naturally dampen if no new forces are applied, a part of the present weight shifting on the planet must contribute to maintaining the wobble. Only the force couples in excess of those needed to maintain the wobble are available to increase the wobble amplitude or drive the axis drift. If the vector sum of all force couples is less than that needed to maintain the wobble amplitude, some dampening will occur.

The question remains, how do major variations in the kinetic loading change the cycle period? We do know that during a period between 1925-1928, when the amplitude of the wobble was at a minimum, the phase of the wobble shifted about 180 degrees, as shown in Fig. 4.1. We also know that the frequency of amplitude peaks, as shown in Figs. 4.5-4.8, are not constant. Since the values that determine the above plots are derived by averaging many observations, the cycle variation cannot be entirely attributed to the limited precision of the observation data. It is also possible that the present 14.29-month cycle is a product of multiple cycles with their combined phases. It will probably take several years of polar movement observations, using the most precise measurements, to completely isolate the contributing multiple cycles. Specially difficult are the smaller variations that are now judged to be noise or random perturbations.

Our Planet's River Erosion Patterns

It is easy to overlook a major contributor to the dynamic balance change by assuming that the stream erosion patterns around the earth tend to be balanced. The effect of weight shifts resulting from stream erosion is determined by several factors starting with the land-ocean distribution and the shapes of their drainage basins. Precipitation rate, type of precipitation, stream gradient, ground cover, and crustal composition all play a part in determining the amount of sediment a stream transports. Should the climate of the Sahara Region suddenly switch from arid to very wet, erosion would be extremely heavy until vegetation takes root. During high water, the faster running streams carry a proportionally heavier load of sediment and nudge the larger boulders downstream.

A world globe is an excellent tool for making a quick assessment of the additive and counter weight shifts created by the major river basins. For this analysis, we are primarily interested in the north and south flowing rivers. They create force couples (see Appendix C for a discussion of force couples) that act on the earth's axis as a result of the sediments they transport. It is necessary to look at the weight shifts one meridian great circle plane at a time, to see if the weight shifts are additive or compensatory for the force couples they produce. The influence of any weight shift, relative to a selected reference plane (shift meridian), is a function of its distance from the shift meridian, decreasing to zero at the neutral meridians.

Northern half of Western Hemisphere

If we look directly at the 90-degree west meridian of the Northern Hemisphere on a world globe, the Mississippi River and its delta are front and center. The Mississippi River and its major tributaries; the Ohio, Missouri, Arkansas, Tennessee, and Red Rivers drain a vast area of the Central United States from the eastern slope of the Rockies to the western slope of the Appalachians. The primary flow direction is to the south.

To the left of the 90-degree west meridian we can see the Rio Grande and Colorado Rivers, also with a primarily south flow as they drain the southern and southwestern Rockies. Both of these river basins are close enough to the 90-degree west meridian to contribute a significant additive force couple. Further to the left are the Columbia and Snake River Basins that drain the northern Rockies to the west. Since they constitute primarily a west flowing drainage they have a limited contribution to the north-south force couples. They are also a little further from our reference meridian. In California's central valleys, the Sacramento and San Joaquin River Basins drain in opposite directions to effectively offset their force couple contribution. The Great Basin of Nevada, Utah, and Southern California is an arid self-contained area with no major river drainage basins.

To the right of the 90-degree west meridian, we find the Tombigbee, Alabama, and Chattahoochee Rivers draining the southern end of the Appalachians. Flowing south they contribute to the Mississippi force couple. The eastern slopes of the Appalachians are drained to the east by a number of relatively short rivers, such as the Potomac and Savannah. They provide a minimal contribution to the north-south force couple. Further north, the Hudson drains to the south, but is making only a limited contribution because its drainage basin is relatively small. The Saint Lawrence River flows to the northeast, draining the Great Lakes Region. The lakes, however, serve as settling basins to dramatically reduce the sediment load that is carried out to the Saint Lawrence Delta.

In Canada, the eastern slope of the Rockies and the Canadian Shield are drained by a series of rivers flowing to the east and north. Although the area is vast in size, two factors greatly reduce the volume of sediments transported. First, the many lakes in the area serve as settling basins to reduce the sediment load downstream. Second, the Canadian Shield is a low relief, hard igneous basement rock surface that is much less susceptible to erosion than the softer sedimentary beds found in most other drainage basins. Consequently, this potential counter force couple is reduced.

The west side of the Canadian Rockies is drained by a series of short west flowing streams. In the northwest corner of Canada, the Mackenzie River Basin flows to the North. The effect of its force couple is counter to the Mississippi, but its effect is limited. It is a relatively small basin and is quite a distance from our reference meridian. In Alaska, the west flowing Yukon is so far from the 90-degree meridian that it is just visible.

Since Central America is a relatively narrow strip of land, its drainage is to the east and west with short streams that contribute little to the north-south force couple. From this quick assessment, it is easy to see that the dominant force couple provided by the drainage basins of the North American Continent comes from the south flowing rivers, primarily in the United States.

Southern half of Western Hemisphere

We next rotate our globe, still looking directly at the 90-degree west meridian. This time we view the Southern Hemisphere, and find mostly the Pacific Ocean at the center and left side of the globe. To the right we see the Pacific side of the Andes being drained by relatively short streams flowing to the west. East of the Andes, the east-flowing Orinoco and Amazon Rivers contribute little to the north-south force couple. Exceptions are the north-flowing Mamore, Juruena, Araguaia, and Madeira River tributaries of the Amazon. In southern South America, the south-flowing Paraguay, Uruguay, and Parana Rivers drain most of southeastern South America. They produce a sizable north-south force couple that is additive to the North American force couple noted above.

Northern half of Eastern Hemisphere

Next we can rotate the globe so we can view the 90-degree east meridian in the Northern Hemisphere. Here we find a predominance of north-flowing rivers in northern Asia, including: the Ob, Yenisey, and Lena Rivers that drain lands north of the Himalayas. A north-flowing river on this side of the globe will produce an additive north-south force couple to the South flowing river majority we found in North and South America. The south-flowing Volga provides a counter force couple. Just in sight on the west horizon of the globe is the south-flowing Dnieper of western Asia, and the north-flowing Nile of Africa. Their influence on the 90-degree east meridian force couple is very limited, as indicated by their lack of prominence from our vantage point above the globe.

Eastern Asia is drained by the Amur, Yellow, Yangtze and smaller streams that flow to the east. The south-flowing Mekong provides the major counter force couple from southeast Asia. In southern Asia, the Ganges flows to the southeast, whilst the Indus flows to the southwest. They also provide a counter force couple, but their influence is reduced by their relatively short lengths.

Southern Half of Eastern Hemisphere

When we rotate the globe to view the Southern Hemisphere over the 90-degree east meridian, we see mostly Indian Ocean with Australia to the right. Since Australia has no major river drainage systems comparable to the other continents, its contribution is limited. Africa is on the extreme left as we view the globe, again with little to offer in the way of an effective north-south force couple.

Other considerations

There are two additional points to be considered when discussing stream and river erosion. First, when sediments are deposited in the oceans at the river deltas, the effective increase in deposited weight is the sediment weight less the weight of the displaced water. The displaced waters, in turn, redistribute around the globe as a function of the ocean pattern (heavily weighted toward the equator and Southern Hemisphere). Since river sediments displace ocean water, the effective weight and resulting force couple are slightly lower than those calculated for the full load of river sediments. Second, when sediments are transported toward the pole, their effective centrifugal force is reduced because of the shorter radius of rotation. Conversely, sediments moved toward the equator have a greater spin radius and apply a larger centrifugal force torque on the axis.

Ocean waters distributed around the earth create their counter balances. Isostatic balance redistributes the displaced magma seas under the river deltas, but not in the same time or with the same freedom of movement as for the ocean waters. The force couples for sediments deposited in flood plains are influenced by the difference in spin radius, but are not reduced by displaced waters. The weight shift resulting from the incremental sediment movement in the erosion process, like that in a pipeline, can be ignored, except for the overall input and discharge of weights.

In summary, the controlling force couples for the 90-degree east and west meridians appear to be the south flowing rivers of the Americas, and the north flowing rivers of Asia. These force couples are counterclockwise as viewed from above Greenwich Meridian (0-degrees longitude), and are consistent in direction with the present drift of the North Pole south towards New England.

Other hemispheric divisions

Although the 90th meridians are probably not the true centroid of the identified stream erosion force couples, they appear to be fairly close. The axis drift towards 70 degrees west longitude, as noted in chapter 4, indicates that the meridian pair 70 degree west and 110 degree east represents the centroid of all force couples for the period pole motion data have been collected.

We apply the same analysis used above for the 0-180, 45 west-135 east, and 45 east-135 west meridian pairs. They each fail to reflect a similar predominance of erosion weight shifting from stream erosion, as noted for 90 west and east meridians detailed above.

Crustal Massaging

Crustal massaging, resulting from magma tide responses to the axis wobble, furnishes a clue to the causes of earthquakes and volcanic eruptions. Consider an axis movement that has the poles tracing a circular path over a 14.29 month period. If we assume for the moment, that the center of the circular trace is the geodetic reference system pole, the meridian passing through an instantaneous pole position is a negative quadrant shift meridian. The shift meridian in the negative quadrant defines the trough of a magma tide as the pole circles the earth. The opposite shift meridian, in the positive quadrant, defines the crest of the magma tide, also circling our planet.

Axis wobble tide amplitude

The diameter of the axis wobble pole trace reached about 0.73 seconds of arc in 1951, as shown in Figs. 4.6 and 4.9. The influence of such an axis shift on the geoid can be easily estimated by proportionally reducing the values along the shift meridian obtained from Table B 1 (Appendix B). The resulting change in the shift meridian geoid, at 5-degree intervals for an 0.73" axis shift, is given in Table B 6 (Appendix B).

The pole's circular path therefore creates a tidal response in the geoid. The tide amplitude (from trough to crest) is about 3 inches at 45 degrees latitude, and decreases to zero at the poles and equator. The tide period is approximately 14.29 months with the trough trailing the crest by approximately 7.145 months. If the crust was rigid, this 3-inch geoid pulse would be reflected in records of tide gauges positioned at or near 45 degrees latitude as the ocean waters respond.

Crustal response to axis wobble tide

As discussed in previous chapters, the same centrifugal forces that create a geoid tide due to axis wobble, also act on the asthenosphere to create a companion magma tide. Magma tides created by axis wobble differ from ocean tides in two critical ways. First, the magma is confined by the crust. Second, the magma viscosity reduces its response time. Pressure on the crust increases as the crest of the magma tide passes. Pressure is reduced as the trough passes. The crust will flex in response to the ever changing magma pressure, depending on its tensile strength and elasticity. If the crust responds 100 percent to the magma tide, the tide gauges will not detect a geoid tide since the ocean waters and crust are oscillating in phase. A more reasonable scenario is that the crust responds excessively where the crust is weakest (plate edges, interior plate fracture zones, and where the crust is thinner, such as the Yellowstone Hot Spot and ocean basins), and remains less responsive where the crust is strongest.

Increased pressure on the lithosphere, as the magma tide crest passes, will push magma up into crustal fractures and weaknesses. Dike and sill intrusions occur first. Any new crustal ruptures, including the extension of existing fractures, are recorded by seismic instruments as earth tremors. Intrusions into batholiths, laccoliths, and magma chambers under volcanoes, trigger more earthquake vibrations. If higher magma pressure is relieved by local intrusions, the tide gauge station (not affected by the intrusions) will record the geoid pulse. An over response of the uplifted crust will not only cancel the effect of an elevated geoid, but indicate a local drop in sea level. Considering the small tide amplitude generated by the axis wobble, the effect on the ocean tide on tide gauges (regardless of scenario used) can easily be overlooked as noise in the tidal records. If all upward and downward warping occurs in the thinner oceanic crust, sea levels will remain essentially unchanged as measured by tide gauges.

Axis wobble tide and Mount St. Helens makes for an interesting speculation. The volcanic eruption of Mount St. Helens can be explained as a possible relief valve for the pressure increase caused by axis wobble induced magma tide, if the crust does not flex under the increased magma pressure. Mount St. Helens, at 46 degrees latitude, is in the path of the largest amplitude of magma tide responding to the axis wobble. The eruption on 18 May 1980 occurred during a period of decreasing magma pressure (about 5 months after the peak pressure applied by the axis wobble). This could represent the lag time response of the magma crest or simply the eventual discharge of magma that moved up into the magma chamber to weaken the volcanoes plug.

Consider magma pressure sufficient to lift the Washington and Oregon Region about three inches. If released through a single orifice, that is more than enough pressure on the crust to blow the top off Mount St. Helens, provided the pressure is not released by crustal flexing.

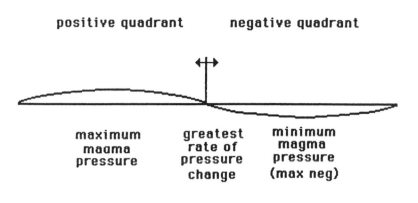

Fig. 5.2 Change in pressure on crust due to Chandler wobble generated magma wave.

As the magma tide circles the earth on a 14.29 month cycle, the crust is being continually massaged. Even when the crust flexes in response to the magma tide, but does not rupture, it is continually weakened. An eventual rupture of the crust can be compared to a piece of wire or metal sheet that breaks after continued flexing.

We can analyze the pressure change exerted by the magma tide over its 14.29 month cycle. It is obvious that the maximum pressure occurs when the tide crest passes (the apex of the tide's sine wave), as illustrated in Fig. 5.2. The minimum pressure (or maximum negative pressure) occurs as the trough passes. The period of greatest rate of change in pressure, however, occurs midway between the crest and trough of the wave. Crustal failure, such as volcanic eruptions and earthquakes would be expected to occur at or near the pressure extremes. Failure could occur at any point in the tide cycle, as the crust is adjusting to differential pressures.

Although the above statements about a positive magma tide responding to the axis wobble are true under special conditions, we need to be aware of other influencing factors. First, we have no way of establishing an absolute origin for positive and negative quadrants. We can only say that a specific axis shift will add magma pressure in two quadrants, while reducing it in the opposite two quadrants. The positive quadrant of the axis wobble cycle may increase existing positive magma pressure or reduce existing negative pressure. Second, because of the high viscosity of the magma seas and the inertia of the crust to respond to pressure variations, the axis wobble tide may pass before the crust responds.

In the Mount St. Helens scenario, the longer term drift toward about 70 degrees west longitude provides a negative quadrant effect on Washington State. If axis drift was the dominant force, Mount St. Helens should never have erupted. Other contributors to the magma pressure under the region include the earth tides generated by the moon and sun's gravitational pull. Variation in magma pressure, resulting from tectonic plate movements (specially when tectonic plates are driven down like a hydraulic plunger into the magma sea), also enters into the equation. In the case of Mount St. Helens, a slab of the Juan De Fuca is being driven down along the subduction zone along the Pacific coast from Oregon northward as the North America Plate moves westward. This adds to the magma pressure in the region and may be the primary force responsible for the Mount St. Helens eruption.

Crustal response to moon and sun tides

Lunarand solar tides are well understood phenomena. They result from variations in the gravitational attraction of the moon and sun, depending on the relative orientation and position of the three bodies. Earth tides are the minute vertical undulations in the earth's crust responding to the same gravitational forces that creates the high and low ocean tides that circle the globe. Earth tides, however, cannot be strictly crustal phenomena any more than we would expect a feather to be lifted off the crust's surface as the moon and sun pass overhead. Earth tides, like ocean tides, must arise from the the oscillation of the supporting magma seas to alternately lift and lower the crustal boat. The gravitational pull by the moon and sun reduces the crust weight. The constant pressure in the magma seas generates an earth tide, even if magma viscosity prevents a geoid-like response.

We have seen that the axis wobble has two major cycles with the annual tide circling the planet every 12-months, and the Chandler tide about every 14.29-months. We have noted that the Northern Hemisphere crest of the wobble tide (defined by the shift meridian) combine with the trough of the tide in the Southern Hemisphere. The crest and trough of each hemisphere are 180 degrees out of phase. Gravitational tides simply add several new cycles into the equation because of the earth's spin rate, moon's elliptical orbit around the earth, and the earth's elliptical orbit around the sun.

The algebraic sum of all the contributing cycles determines the magnitude of the geoid pulse and magma tides. When the sum of the individual cycles tend to offset each other their influence is at a minimum; it is at maximum when all are in phase. The role of oceans and earth tides are discussed in greater detail in Appendix E.

Axis Wobble Correlation With Earthquakes

One suggestion is that earthquakes may cause the axis wobble. Efforts to correlate the occurrence of large earthquakes with the axis wobble, however, have not supported this hypothesis. Based on the Dynamic Axis Theory, it appears that the correlation should be reversed. If magma sea tides generated by axis wobble are massaging the crust, the massaging can account for crustal gymnastics. Due to a lag in the response to magma tides, earthquakes and volcanoes should follow, not lead. Earthquakes, of course, are simply the physical evidence that crustal displacements occur.

The sum of magma sea responses to the moon's gravitational pull, the sun's gravitational pull, axis wobble, axis drift, and changes in length of a day determines the relative stresses applied to the crust. A computer program is required to correlate magma tide induced stresses with earthquake occurrences. The program needs to monitor axis wobble pole positions and gravitational driven earth waves with earthquake and volcanic eruption events, both as isolated and joint triggers. The Dynamic Axis Theory predicts that we should find a correlation between crustal movements and effective magma tide. Failure to find a correlation would be cause for reviewing the theory's validity.

Earthquakes as a function of latitude

If we only consider the crustal stretching caused by axis shifts (see Tables B 1 and B 2, Appendix B), the greatest crustal stress occurs near 45 degrees north and south latitude. The stress decreases to zero at the poles and equator. The sum of crustal stress from lunar tides, solar tides, and changes in length of a day is largest at or near the equator. Combining the centrifugal, gravitational, and length of day induced stresses moves the maximum crustal stress toward the equator as illustrated in Fig. E 4 (Appendix E).

The widths of the transverse fracture zones associated with the midocean ridges indicate the crustal stress. They are in general agreement with the above analysis. The maximum stress (maximum width of transverse fracture zones) appears to occur between 20-30 degrees latitude in both hemispheres of the Atlantic and Pacific Oceans.

At first glance, it seems reasonable to assume there should be some correlation between the greatest stresses and the frequency of earthquake occurrences. Expectations and reality do not always agree. The histogram in Fig. 5.3 shows the frequency of 6.0 and stronger earthquakes from 1948-1986, grouped in 5 degree bands of latitude. Clearly, the distribution is not uniform by latitude. The distribution also does not fit the stress pattern shown in Fig. E 4 (Appendix E). The highest frequency of earthquake activity

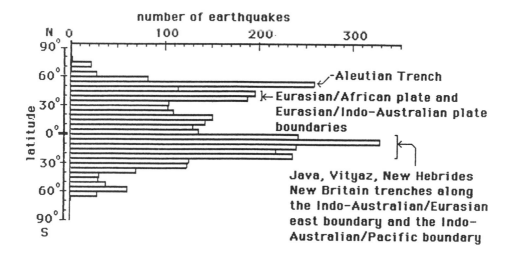

Fig. 5.3 Earthquake frequency (6.0+ magnitude earthquakes for 1948-1986) as a function of latitude. A = Aleutian Trench; B = Eurasian/African plate and Eurasian/Indo-Australian plate boundaries; C = Java, Vityaz, New Hebrides, New Gritain trenches along the Indo-Australian/Eurasian eastern boundary, and the Indo-Australian/Pacific boundary.

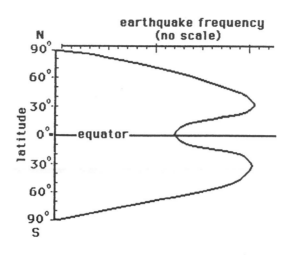

Fig. 5.4 Anticipated earthquake frequency pattern by latitude. Trench between 20 and 35 degree south latitude. The high count between 10 and 20 degrees north latitude can partially be explained by the fact that the Middle American Trench is aligned in a more northwest to southeast direction from Manzanillo to Panama. Considering the movement direction of the Pacific Plate and the extra length of the contact surface, higher earthquake frequencies in this area are not surprising.

occurs between 5 and 10 degrees latitude in the Southern Hemisphere. Earthquake frequency is high for the band of latitude from the equator to 30 degrees south. In the Northern Hemisphere there are two bands of high earthquake activity: one between 15 and 25 degrees latitude, and the second between 50 and 55 degrees latitude. This is not exactly what Fig. 5.4 suggests, but why?

Plate edges could provide the answer. When the tectonic plates dance to the tune of magma massaging, the stress within a rigid plate transfers to the plate edge where most ruptures occur.

For the eastern edge of the Pacific Plate, the highest concentrations of earthquakes occurs along the Middle America Trench from 10-20 degrees north latitude and the Peru-Chili Trench from 20-35 degree south latitude. The high count from 10-20 degrees north latitude can partially be explained by the Middle American Trench's alignment in a more northwest to southeast direction from Manzanillo to Panama. Considering the movement direction of the Pacific Plate and the extra length of the contact surface, more earthquakes in this area are not surprising.

On the western side of the Pacific, earthquakes contributed by the Java, New Britain, and Vityaz trenches beef up the count for the zone from the equator and 10 degrees south latitude. Although no trench exists between the Java and New Britain trenches, including New Guinea, the frequency of earthquakes is very high. This could indicate the existence of an overthrust belt as opposed to a subduction zone. In the west central Pacific, a complex system involving the Indo-Australian, Eurasian, Caroline, Fiji, and Philippine plates accounts for a concentration of earthquakes in low latitudes.

The most dramatic example of earthquake frequency shifted to a plate edge is the Aleutian Trench where the Pacific Plate is subducting under the Eurasian Plate. The east-west alignment of the Aleutian Trench accounts for the high occurrence anomaly at 50 to 55 degrees north latitude.

The south edge of the Eurasian Plate from the Himalayas to the Alps straddles 30 degree latitude. Strong and frequent earthquakes occur along the southern border of the Eurasian Plate from the Himalayas and Alps, mostly just north of 30 degree north latitude.

Spreading zones represent very active crustal gymnastics. An interesting sidelight to the distribution of large earthquakes, however, is that relatively few happen along the spreading midocean ridges. The reason relates to the fact that large earthquakes occur where friction locks the plates together to buildup the stress before they rupture. When two plates are being pulled apart, the required friction lock is missing after the initial rupture. Since the data on earthquake occurrence used for this study involved only the 6.0 and larger shocks, smaller earthquakes along the midocean ridge are excluded from the count.

Transverse fracture zones associated with the midocean ridge provide another puzzle because of their dearth of strong earthquakes. The answer appears to be that the transverse fractures are nearly parallel to the shearing forces that are pulling the plates apart at the midocean ridges. Without a sizable force acting perpendicular to the transverse faults, they are incapable of building friction locks, such as those formed along the San Andreas fault where the North American Plate is driven into the Pacific Plate. Also the transverse zone fractures are nearly straight, unlike the crooked alignment of the San Andreas fault that adds to its friction lock. Consequently, the earthquake activity associated with the transverse fracture zones is expected to have a high frequency, but with lower intensity.

The Earth's Tailor

Nature has the same problems of continually remodeling its body suit (crust) to fit the asthenosphere as a tailor who trys to remodel a body suit for a girl as she grows into a young woman. For illustration purposes, consider a 12 year old girl, 5'-0" tall, weighing 115 pounds. She received a very special body suit for her birthday. The material was so beautiful that she had it altered as she grew into a woman. Fortunately, in the original tailoring of the body suit, extra material was left at the seams to allow for the alterations. As she became a 5'-9", 115 pounds young lady, the suit was altered regularly to maintain its perfect fit. Because of her growth in height, seams were let out or gussets were added to accommodate her taller body. This would compare with the midocean ridges of our planet where new crust is formed to fill the gap. Since our young lady maintained her weight while she grew taller, she also reduced her waistline, making it necessary for the tailor to use pleats to take up the excess material. The earth's tailor uses synclines, anticlines, subducting plates, and overthrust belts to take up its excess crust. Other, more subtle changes of the girl's body shape each require some local alterations to maintain the perfect fit. These are comparable to batholiths.

Conveyor Belt Theory

Before we continue our discussion of the Chandler wobble and axis drift, which is fundamental to the Dynamic Axis Theory, we need to analyze the notions of the theory that magma currents provide a conveyor belt driver for tectonic plates.

South American Plate

The conveyor belt scenario fits the South American Continent quite well. We assume upwelling hot magma along the Mid-Atlantic Ridge flows to the west and drives the South American Plate over the subducting Nazca Plate along the Peru-Chili Trench. A companion magma undercurrent flows in the reverse direction to complete the loop. The indicated current magnitude, however, is inconsistent with the multi-cellular circulation patterns of fluids heated in a pan, where the cell size is related to the depth of the circulating fluid. The Conveyor Belt Theory implies that the width of the magma current carrying South America must extend from the north end of the continent to beyond the southern tip. The Conveyor Belt Theory, for driving the South American Plate, calls for a circulating cell 6,200 miles long (east-west direction). Lab tests on heated fluids only support about 400 mile long cells, if circulation is confined to the upper mantle, or about 1,700 miles, if the entire mantle circulates as a unit.

Furthermore, the proposed magma currents are not consistent with our knowledge of ocean currents. Using the conveyor belt logic, the ocean currents heated along the midocean ridges would rise and split to flow away from the ridges in both directions. Instead, ocean currents have a horizontal circulation pattern. They are only slightly influenced by the linear heat source provided by upwelling magma along the midocean ridges of the world.

Superimposed on the ocean's horizontal circulation pattern are vertical circulating eddies. The ocean eddies, however, are not even remotely similar to the proposed magma circulating patterns. Using the conveyor belt concept, both atmospheric and ocean currents heated in the tropics and cooled in the polar regions should circulate as north-south vertical loops.

Researchers at the Department of Earth and Space Science, University of California in Los Angeles, and the Space and Sciences Division, Los Alamos National Laboratory in Los Alamos, New Mexico, have made efforts to model the mantle dynamics. In establishing the model parameters, they used three assumptions: the mantle was heated from below, from within, and both from below and within. In all cases, they obtained planar sheet downwelling and cylindrical plume upwelling. The downwelling fits the subduction end of the Conveyor Belt Theory, but there is no indication of a corresponding planar sheet upwelling at the midocean ridge.

Examination of other tectonic plates generates more questions about the conveyor belt theory.

Pacific Plate

The huge Pacific Plate brings up some additional questions about the hypothesis of conveyor belt currents other than the size. The midocean ridge across the southern Pacific, combined with the ocean trenches along the Aleutians and northwestern Pacific, indicate that the plate must be driven by a south to north conveyor belt current. The north-south midocean ridge in the southeastern Pacific, and the series of ocean trenches along the western Pacific indicate a more westerly flowing conveyor belt current. The Pacific Plate's actual movement appears to be north-northwest as it slides past the North American Plate along the San Andreas fault zone—a reasonable compromise if two independent forces are being applied to the plate.

North American Plate

The North American Plate moves west, just as the South American Plate, away from the Mid-Atlantic ridge. On the west coast, from northern California to Alaska, it is being driven over the Juan De Fuca Plate. To the south, Mexico is driven over the Cocos Plate. In between, it appears that the North American Plate has completely overridden the ancestral midocean ridge, of the Juan De Fuca and Cocoa plates—probably a single large plate at one time. The inland trace of the San Andreas fault fits the projected alignment (extension) of the Pacific Ridge quite well.

African Plate

Africa raises the biggest question in that the southern part of the Mid-Atlantic Ridge defines the west edge, whereas the midocean ridge in the Indian Ocean defines the eastern edge. Both are spreading zones with no indication of a companion subduction zone or plates in collision. In fact, a new rift zone is reported to be opening in the eastern part of the African Continent. The portion of the African Plate east of the rift valley is referred to as the Somali Plate, however, its boundary is not completely defined at this time. All ridges and rift valleys, being aligned generally north-south, indicate any conveyor belt currents must be flowing east and west. Africa, however, is also assumed to be in collision with the Eurasian Plate and is responsible for the formation of the Alps. This could still be true if the African Plate has little northward movement and the driving force associated with the collision is supplied by the Eurasian Plate.

Eurasian Plate

The Eurasian Plate does not appear to have the necessary plate spreading along the North-Atlantic Ridge and its extension into the Arctic Ocean to account for all the mountain building along its southern edge. The small magnitude of apparent plate spreading in the Arctic indicates that the majority of the thrust responsible for the Alps and Himalayas must come from the north bound African and Indo-Australian plates. However, we just noted that there is little indication that a conveyor belt is driving the African Plate to the North. The crustal rupture that connects the Mid-Atlantic Ridge to the Mid-Indian Ridge below Africa appears to be more of a transverse fault slippage, not a plate spreading zone. For Africa to provide the driving force for the Africa-Eurasia collision, the plate edge between Africa and Antarctica would have to be a spreading zone.

The large Eurasian Plate may be the most stable plate with the interaction between it and adjacent plates primarily dependent on adjacent plates. This means the subduction zone along the series of deep ocean trenches of the western Pacific is attributable primarily to the Pacific Plate movement in collision with a nearly fixed Eurasian Plate. We just noted, however, that the Pacific Plate is supposed to be move north-northwest, or nearly parallel to some of the Pacific Trenches. Considering the relative size of the Eurasian Plate, it is easy to assume its movements are slight. This explanation of the Eurasian Plate stability, however, is inconsistent with substantial movements of the huge Pacific Plate.

Indo-Australian Plate

The spreading midocean ridge of the Indian Ocean sends the Indo-Australian Plate northward. The west half of the plate, in collision with the Eurasian Plate, creates the Himalayas. The east half of the plate, also in collision with Eurasian Plate, is defined by the Java Trench. This region has many complexities, but most appear to be consistent with the northward plate movement regardless of the driving force.

Antarctica Plate

Antarctica has a midocean ridge pushing it away from Australia and the South Pacific, but it has only the relatively small South Sandwich Trench off the east coast of South America to absorb the thrust. This area of subduction is associated with the small Scotia Plate, as opposed to being directly related to the Antarctica Plate.

Small Plates

Several small plates have been identified, specially on both sides of the Central American Peninsula, along the west coast of North America, in the Mediterranean, and in the Near East. It is easier to visualize the proposed magma current conveyor belt patterns as a driving force for the smaller plates than for the very large tectonic plates.

Tomographic Maps

The 3-D tomographic maps that show rising hot magma and sinking cold magma within our planet raise some interesting question. The conveyor belt theory calls for rising hot magma along the midocean ridges, and cooled subducting magma on the opposite side of the plates. For the North American Plate, the size of the hot magma along the Mid-Atlantic Ridge is even less than the hot magma under the western United States where cool magma should begin its downwelling. The cool magma is shown under the eastern two-thirds of Canada and extends down into the eastern half of the United States where the warm magma should be flowing westward under the crust. Hot rising magma is also shown in the deep ocean trench region from Japan to New Guinea, as well as the eastern part of the Aleutian Trench subduction zone.

Gravitational variations

Variations in the earth's gravity give strong indication that the mantle is heterogeneous with high and low density circulating cells. If such cells exist, they should flow with a prevailing current pattern of the asthenosphere similar to the atmospheric high and low cells. In time, the undulations of the geoid that reflect a gravity variation would drift to occupy a different region of the globe. This would cause some small changes in the ocean beaches by elevating some while lowering others. The changes in sea level proceed extremely slowly and would not account for distinct marine terraces. Mapping the subtle marine terrace changes as a function of time, however, could contribute to our knowledge of magma current flow. Also relevant in this context is that the movement of high and low magma density regions in the asthenosphere contributes to changes in the centrifugal forces and the force couples acting on the earth axis. Due to the high viscosity of the magma seas, the drift rate is very slow and could be a major factor in the axis drift. A slow accumulation of centrifugal force imbalances may be released as a sudden surge of the earth axis position, similar to stresses that buildup in an earthquake fault zone before rupturing. The indications of axis surges resulting in major cataclysmic events over geologic history will be discussed in chapter 7.

6   CATACLYSMAL UNIFORMITARIAN EVOLUTION

On the surface, cataclysmal and uniformitarian evolution appears to be an oxymoron. The Cataclysmal Theory proposes that the shaping of the earth's crust is caused by sudden and violent upheavals. Cataclysmic change is often associated with a deluge of myths and legends, such as the lost cultures of Atlantis. It therefore automatically alienates some people who refuse to consider anything with religious or mythical overtones. Uniformitarianism, on the other hand, suggests a gradual and often undetectable rate of change. It does not, however, rule out minor local catastrophes. Whereas it has taken millions of years to push up the Rocky Mountains and for erosion forces to carve the majestic Grand Canyon, the rate of change has been very non-uniform.

Cataclysmic in this writing refers to occasional violent surges of change as being a natural part of the overall evolutionary change. Throughout the natural sciences we find numerous examples of short periods of very rapid or violent change that interrupt the perceived normal evolution. As with all scientific studies, the clue can often be found in the natural process of other scientific events. A review of some of these small and medium size cataclysmic events will help us understand how Mother Nature thinks. We will know what to look for in the long range picture of geologic changes. There are many more changes that indicate past global cataclysmic surges than is documented in history.

Cataclysmic Events in Today's Climates

Each year climate extremes occur frequently around the globe to physically alter our planet, kill thousands of people, and cause millions of dollars in property damage. Wind is a common ingredient in many of the cataclysmic climate events.

Winds

Air currents that are normally classified as a breeze or light wind may gust as destructive forces for short periods of time.

A hurricane's cyclonic winds, born in the subtropical ocean waters, have a life span of a few days at the most. The combination of strong wind, heavy rain, and high tides along coast lines can add up to a destructive force. Each year many lives and millions of dollars in losses are attributed to hurricanes (typhoons in the China Sea) around the world.

Meteorologists can now recognize the embryo that has the potential to develop into a hurricane. Heat provides the driving force that creates a low pressure circulation cell and adds moisture by evaporation from the ocean surface. When the warm moist air of a hurricane encounters cool air fronts, the probability of volatile air currents triggering heavy rains and tornadoes increases. The hurricane intensity quickly diminishes as the sea-born storm moves in over land areas, or over cooler ocean waters.

Meteorologists, armed with data from satellite weather observation stations, planes that fly into the eye of the storm, and ships near the tropical depression, plot the path and intensity of each low pressure cell. Officially, they classify as tropical storms until the winds reach hurricane strength of 75 miles per hour. High speed electronic calculations and communications have eliminated much of the guesswork of predicting hurricanes, but the exact path and wind velocities are never completely known in advance. Enough, however, is known to save thousands of lives.

Over the last 80 years the Atlantic has spawned an average of less than 5 hurricanes per year. Many more hurricanes or typhoons are spawned in the Pacific and Indian Oceans. Only when a long list of variables responsible for determining our weather combines in just the right mix, do we experience Nature's weather extremes. When we consider the area in the direct path of hurricanes, only a small percentage of world land areas and population centers are directly affected by their most destructive forces. That statement is of little comfort to the people in southern Florida. They bore the blunt of hurricane Andrew in the summer of 1992, and were then hit again by the winds and tornadoes of the "Storm of the Century" that swept up the east Coast of United States and Canada in March of 1993.

Tornadoes represent even more destructive winds and usually last only for a minutes. The conditions that spawn tornadoes usually emerge and disappear within hours. The occurrences of tornadoes are so common in the midlands of the United States that the region is also known as "tornado alley." Although their occurrences are common, the land surface directly affected by tornadoes, like hurricanes, represent only a minute fraction of the earth's surface area. The odds of an individual building being hit is remote—if it was otherwise, insurance would be nearly impossible to obtain.

As with hurricanes, meteorologists now understand enough about the physical mechanics of tornadoes to provide advanced warning of a potential outbreak of violent weather. They can, however, not predict where, when, or with certainty that a tornado will strike. Although structures cannot be protected, severe weather warnings allow people to take cover in safe shelters. It is still a problem to notify the public on short notice and to take prudent precautions. Curiosity will always entice some to take a chance by watching the funnel clouds form.

Other destructive winds, referred to as wind shear, microbursts, and wind rotors are familiar terms for the dangers they pose, specially to aircraft during take-off and landing. Local high velocity wind gusts present a greater risk to aircraft when the downdraft or turbulence occurs close to the ground where the pilot has less time and space to recover. Even horizontal gusts of straight winds, created by marked differences between adjacent high and low atmospheric pressure cells, can wreak havoc with buildings, high profile vehicles, and anything not firmly attached to the ground.

Hail is another by-product of the volatile summer storms—the same weather systems that create flash floods, electrical storms, microbursts, and tornadoes. We accept these surges as a natural part of our climate cycles. However, the term "cataclysmic" is definitely appropriate for people directly affected by their destructive forces.

Flash floods

Flash floods are part of the normal hydrological cycle that circulates approximately 10 million billion (10^{16} or 10 followed by 16 zeros) gallons of water each year. With 97 percent of all the earth's waters in the ocean, its evaporation accounts for about 86 percent of the moisture source for the hydrological cycle. Evaporation from rivers, lakes, and streams provides another 12 percent. Transpiration from plants and trees produces the remaining 2 percent. In spite of the impressive numbers, the active hydrologic cycle involves only about 0.005 percent of the earth's 326 million cubic miles of water. Except for temporary spot accumulations of water (such as in lakes, aquifers, winter snows, and glaciers), the weight shifts on our planet resulting from the hydrologic cycle can be considered in near balance.

Each spring, Nature's surge of snow melt fills rivers over their banks in a natural process that flushes the valleys of some of their dead vegetation and animal carcasses. High waters boil with sediments washed from the mountain sides. The erosion process continues to carve out the majestic alpine topography. Flood waters spill out of their banks, carve new channels, and invade lowlands.

Big Thompson flood. For most days of the year the residents and visitors of the Big Thompson Canyon, between Loveland and Estes Park, Colorado, enjoy the picture postcard scenery of the Big Thompson River and its tributaries (including the eastern drainage of Rocky Mountain National Park). Its crystal clear waters serve as a magnet to the avid fisher. Campgrounds fill with picnickers when the parka weather fades in the spring. The canyon highway serves the motoring sightseers that are satisfied with just a passing glimpse at one of Natures wonderlands.

In a matter of minutes on 31 July 1976, the Big Thompson River turned into a raging torrent, eroding away sections of the highway and destroying houses along the river bank. Twelve inches of rain fell over a 4 hour period starting about 6:30 P.M. A flash flood, up to 20 feet deep, hit with such suddenness that 39 lives were lost, including state highway patrolman, Sergeant W. Hugh Purdy, while he was trying to save the lives of others. Three more people remain missing and are assumed dead.

About as quick as it came, the flood waters subsided. They left a path of physical destruction, many long term memory scars for families that had lost relatives or friends, as well as everyone else that was touched by the cataclysmic flash flood. Tons of Rocky Mountain soil were bulldozed just a little further toward the sea.

Ratings are often assigned to flash floods based on their expected frequency. The ratings present more a gut feeling than a scientifically derived value. The Big Thompson flood has variously been referred to as either a 300 or 500 year flood. With dozens of canyons carved out of the Rockies, a 300 year flood can be expected to occur someplace along the Colorado Front Range every few years.

Kansas City flash flood. On 11 September 1977, the weather bureau in Kansas City predicted a 70 per cent chance of thunderstorms for Sunday night and Monday. Just after midnight on Sunday, Kansas City received about 6 inches of rain causing some flooding. Another flash flood watch was raised for Monday afternoon and night. At 5:30 P.M., torrential rains in the 4.5-7.1 inches per hour rate were observed in areas to the north and west of Kansas City. The heavy rains hit the city about 8 o'clock, dumping another 6 to 7 inches of rain. This gave Kansas City two 100 year downpours in less than 24 hours—an estimated once in 1,000 year occurrence.

Johnstown flood. A flash flood caused by heavy rains and the failure of an earth dam crashed down on Johnstown, Pennsylvania on 31 May 1889. The rush of approximately 20 million tons of water caused 2,209 deaths. The destruction wrought by the flood waters has made Johnstown synonymous with floods. Even without the earth dam breaking, Johnstown has suffered through a series of flash floods. As recently as July 1977, a 12 inch rain fall in 7 hours sent the Conemaugh River out of its banks and over the protective barriers built by the Corps of Engineers. The flood was at least the third so-called 200 year flood to hit Johnstown in less than 100 years. This time the death toll was 77.

Shadyside flood. Do not brag about your destructive flood story to the residence of Shadyside, Ohio. For years to come, 14 June 1990 will be remembered as the night 5.5 inches of rain fell between 7:30-11 P.M. to send raging flood waters through their peaceful community.

Flash floods, such as the Big Thompson, Kansas City, Johnstown, and Shadyside occur regularly around the world, each with a different setting and cast of characters; but each leaving their stamp of destruction engraved in the minds of the living victims.

Teton Dam break. Flash floods are not limited to those produced by heavy spring runoff or cloudbursts. On 5 June 1976, the 307 foot high earth fill dam on the Teton River in Idaho ruptured sending its reservoir of waters downstream. Eleven people lost their lives and about 25,000 were left homeless. The storage of waters behind artificial dams constitutes an added spot weight in the earth's dynamic balance. The sudden release of the waters removes the weight to alter the earth's dynamic balance.

Hebgen Earthquake lake. Nature creates many temporary reservoirs by blocking drainage basins with rocks and mud slides. Such a rock slide formed earthquake Lake in Madison Canyon during the Hebgen earthquake in eastern Idaho, just west of Yellowstone National Park. The earthquake responsible for the rock slide that blocked the Madison River hit at 11:37 P.M. on 17 August 1959. The formation of a lake behind the slide created an immediate emergency. Eventually the lake level would rise causing the dam's unconsolidated loose rocks and soils to fail and release a flash flood on the residents down stream. The Army Corps of Engineers cut a 250-foot wide and 14-foot deep channel through the mile and a half long slide to prevent a repeat of the Gros Ventre River slide in 1927. The Gros Ventre's trapped waters broke away two years after its slide blocked the stream. The town of Kelly, Wyoming (on the opposite side of Yellowstone and Grand Teton National Parks from Hebgen) was washed away killing 7 people.

Channeled Scablands floods. It took some Sherlock Holmes quality investigation and keen intuition by geologist J. Harlen Bretz (in 1923), to develop a theory for the geologic formations in the Grand Coulee Dam Region of Washington State, known as the Channeled Scablands. As with most new theories, Bretz was initially met with skepticism when he proposed that the area was carved by a massive flash flood at the end of the Ice Age. He had found evidence of a 2,000 foot high ice dam (the arm of an intercepting transverse valley glacier that had blocked the Clark Fork River in Montana to form a 3,000 square mile lake).

When the ice dam eventually ruptured, approximately 2.8 billion gallons of water per second was released to scour out the Clark Fork, Pend Oreille, and Columbia River Valleys. The force of a massive flood of this magnitude is needed to explain the sculpturing of the scablands. Calculations for the volume of water flowing down the valley were, in part, derived from mile-long giant ripple marks of sand and gravel deposited by the flood. This is the same type of sand ripples that are made by any rush of waters, only much larger in scale.

Terminal and lateral moraines form as glaciers move down a valley. The mass of ice acts like a bulldozer carrying and pushing a bead of rocks and soils. The loose rocks and soils that are deposited at the glaciers farthest advance are called terminal moraines. Soil deposits left along the edge of a valley glacier are called lateral moraines. As the glacier melts, a lake often forms behind either terminal or lateral moraines. Since the moraine dams are composed of an unpacked mixture of rock, sand, and soils, (an extremely poor dam), they create the potential for flash floods when the temporary glacial lake is filled. Once the waters break over or through the moraine, a small opening soon becomes a gaping hole, as in the case of the Teton, Hebgen, and Gros Ventre earth dams and the ice dam on the Clark Fork. The sudden release of waters creates a flash flood that sweeps down stream.

Tropical storms. generate high tides that can slam into coastal regions, causing flash flooding. Not only does the high tide inundate coastal lowlands but wind driven waves crash onto shore with a destructive force. Few natural or man-made barriers can withstand the onslaught of their rushing waters without being altered in some way.

Tsunamis (from the Japanese word for harbor wave) are caused by earthquakes. The wave crest can roll across the oceans to crash into coasts thousands of miles from their origin. In the open sea the tsunamis are only a few feet high. However, when they reach shallow harbors they can be amplified to over a hundred foot wall of water. Crescent City, California was the recipient of a tsunami flood following the 1964 Anchorage Good Friday earthquake. It raced across the Pacific at 400-500 miles per hour. Ten lives were lost in Crescent City despite advanced warnings provided by today's high speed communication. A total of 119 deaths and 94 million dollars damage around the Pacific was attributed to the Anchorage tsunami. A tsunami triggered by the 1960 earthquake off Chili caused 61 deaths and 23 million dollars damage in Hilo, Hawaii alone.

Seasonal flooding

Mississippi River floods are often caused by heavy rains that accelerate the melting of the winter snows in the upper Mississippi and its tributaries. Whenever the timing of flood waters from different tributaries coincides, the lower Mississippi is in danger of extensive flooding. For the past 200 years, the Mississippi has averaged a major flood about every 7 years. One of the preventive measures taken to control the mighty Mississippi is a levee system built by the Corps of Engineers. When the levee holds, the residents bordering the river give a big sigh of relief. When they break, waters race over the adjoining lowlands as a flash flood. Sediments deposited by the flood waters enrich the soils while adding weight to the crust.

When the Mississippi flood hit in 1973, the river crest at Cairo, Illinois was 55.7 feet. This was just 0.7 feet lower than the 1927 flood, and 3.8 feet lower than the 1937 flood. Further south at Vicksburg, the crest of 53.1 compared with 59.9 and 57 feet for the 1927 and 1937 floods, respectively. The 1973 flood waters inundated over 16 million acres. Property losses from the flood in 1973 reached a billion dollars, and nearly 69,000 people were left homeless. Miraculously, only 23 lives were lost. One can only speculate on the losses without the massive flood control projects.

Yellow River floods. In terms of human suffering and loss of life, the above disasters fade in comparison to the historic floods of the Yellow River in China. In 1887 flood waters breached the Yellow River dikes, inundating the flood plains before many farmers and villagers could escape. Estimates range between 900,000-2,500,000 lives lost by flooding, starvation, and disease. An area of fertile farmland the size of Lake Ontario went under water. The Yellow River is known for the heavy load of silt it transports. Of course, the silt deposited in the lowlands near the river attracts the farmer back year after year in spite of the continual flood threat.

Fill in your flood stories. Most every region on earth has had a devastating flood that can be entered into a "can you top this" debate. We all relive the ones that came dangerously close. Destructive floods, like destructive winds, always depend on the additive effect of a series of elements. Just how the contributing factors come into play determines the magnitude of each cataclysmic event. Cataclysm, however, is the right word since flash floods can alter the topography more in a few hours than the changes during the intervening years between floods.

Winter snows and blizzards

Heavy snow storms are winter's counterpart to the summer cloudbursts. Add wind and the animals go for cover just to survive. Large losses of domestic animals often occur in the late spring storms when new born calves and lambs are most vulnerable. As with flash floods, the blizzards and winter storms may be rated by their intensity and frequency. Winds rearrange the winter snows into drifts that extend the spring and summer melting period. In alpine regions, the drifted snows deposit in shaded pockets to create snow fields and valley glaciers that survive the summer melt.

The ice cap build-up that covered much of Canada and northwestern Europe was, undoubtedly, accompanied by many white-out blizzards. It prevailed not only where the ice accumulated, but also over the adjoining land areas. Snow cover forms a reflective surface that reduces the warming effect of the sun and intensifies the winter conditions. When the ice cap thickened and buried all signs of vegetation, the remaining life in the region was reduced to a few hardy species, not unlike that found today in Antarctica and Greenland.

Buffalo, New York in the winter of 1976-1977 accumulated about 3 feet of snow by 28 January from daily snows lasting for 40 days. More snow and 70 mile-per hour winds buried the area under an additional 4-foot blanket, with 30-foot drifts. The 5-day storm took 29 lives and cost approximately $250 million. Warmer spring winds circulating up from the South, quickly eliminated any thoughts that another ice age was descending on the region.

Arctic and Antarctica storms are best known for their winds and whiteout conditions. The low annual precipitation for both areas limits the accumulation but not the drifting that continually rearranges the dry surface snows.

Droughts

Droughts, like winds, floods, and blizzards, are a relative term representing an extreme low in precipitation. The Sahara residents would consider the precipitation of the 1930's drought in Midwestern United States as wet years. Meteorologists statistically determine an average precipitation value by regions and areas. Any period that is substantially below normal represents drought conditions, regardless of how much precipitation is normal for the region.

Just as important to biota is the time and intensity of precipitation. There are several critical periods when crops in dry land farming regions need moisture for the seeds to germinate and progress to maturity. Deficiencies at the wrong time can take a tremendous toll on crop production even when the year totals do not indicate drought conditions. Rain can be lost to run-off when it comes too heavy to soak in, or the lands are saturated. Of course, rains during planting and harvest seasons create their kind of catastrophe for the farmers. Arid deserts in the southwestern United States are noted for extremely heavy rains (often local cloudbursts) followed by years with little or no rain. The beauty of a desert region after a wet winter and spring is just one more of Nature's surprises for those that previously had only seen the desert as arid wasteland.

America's Heartland suffered a devastating drought during the 1930's when its climate cycle turned increasingly hot and dry. Hot dry winds whipped up the soils, loosened by farming, into dust clouds that turned day into night. Lands of America's bread basket, that were adapted to a particular type of cultivation and farming, were unable to support their normal abundant crops. The intensity of the 1930's drought as a cataclysmic event is measured in part by the migration of farmers and business people out of the region. Many left with only a few personal items and their clothes on their backs.

Precipitation well below normal levels created drought conditions in the southeast United States early in the 1980's. In the summer of 1988, lack of moisture fostered talk and fear of a return to the drought conditions of the 1930s for much of America's Heartland. The fires in Yellowstone Park in 1988 were intensified by the below normal precipitation in the region. Many of the areas hit by drought conditions in 1988, however, were back to near normal in 1989.

Regional droughts. Africa has made headlines in recent years, primarily because of the associated human starvation. The toll on wildlife has been equally destructive. No region of the earth is immune from lowered precipitation cycles, including rain forests. The cataclysmic effect of a drought condition is at a maximum where plant and animal life have become dependent on the high precipitation rates.

Temperature extremes

Meteorologists maintain temperature records for a network of recording stations around the world. From the individual records, they generate many statistical reports by station and by region. Weather news broadcasters regularly compare daily temperatures to normal values determined by averaging the temperatures since records have been kept. Weather news headlines are made when the recorded maximum or minimum temperature for any day is exceeded. A string of record lows prompts articles suggesting the pending return of the Ice Age. A string of record highs prompts "I told you so" articles from the proponents of the greenhouse effect.

Many cities will experience some record high and lows sometime during the year. We only need to consider that every city has 730 opportunities to set a temperature high or low record each year, and that a daily temperature plot for any point on earth is patterned by the bell-shaped normal distribution curve. The number of records broken is inversely proportional to the length of available records. The second year of record keeping for any city produces nearly 730 new records with only ties eliminated from the record books.

During the early winter of 1988-1989, the United States had a run of relatively pleasant weather as the jet stream took a generally west to east path across Canada. The jet stream served as a weather steering device, making the division between colder air to the north and warmer air to the south. By flowing nearly straight across Canada, the jet stream prevented the normal mixing between the upper and lower latitudes.

Several record highs were broken in the lower 48 states. While people in the lower 48 states were enjoying an extended fall with above average seasonal temperatures, the heat loss by radiation continued to lower the Arctic temperatures to produce record low temperatures for Alaska. The jet stream finally changed its course, making a big swing down into the lower latitudes to begin the mixing between temperature extremes. Before the air masses could mix and modify the temperatures, the jet stream change brought surges of cold Arctic air south. It set record cold temperatures for many of the same cities in the lower 48 states and southern Canada that experienced record highs only a few days earlier.

Occasionally, the factors controlling climate conditions combine to create a cataclysmic event. If your area is experiencing an extreme in temperatures or precipitation, you can bet that some other area on the globe is setting records for the opposite extreme.

Cataclysmic Events of Today's Lithosphere

Just as climate conditions team up to create cataclysmic storms, the physical forces that act on the planet's crust can team up to produce sudden devastating events.

Earthquakes

The term "earthquake" usually projects an image of physical destruction associated with a near surface rupture of the earth's crust occurring close to an urban setting. Television, newspapers, and magazine photos have brought the destructive power of earthquakes out of the geology classroom, and into the living room. Large earthquakes, as measured by human suffering and economic losses, in San Francisco, Anchorage, Mexico City, Los Angeles, Armenia, Iran, and India have made the headlines in recent years.

Any sudden movement of one block of the crust from an adjacent block sets crustal vibrations in motion. An earthquake is recorded whenever the amplitude of vibrations recorded on the seismograph charts stands out against the background noise (a continual low amplitude vibration signal). Seismologists record the earth's vibrations as the pulse of our planet. By timing the earthquake signals received at three or more seismic stations, seismologists can determine its epicenter location (surface point above earthquake) and focus (actual rupture point below the surface).

The physical intensity of earthquakes is measured on a logarithmic scale, known as the Richter scale. An increase of 1.0 on the Richter scale represents a ten-fold increase in magnitude. Our perception of earthquakes as a cataclysmic event can differ from their physical magnitude, depending on the associated human suffering and economic losses. Compare the Armenia earthquake, at 6.9 magnitude, with the 8.3 earthquake in the South Pacific, south of New Zealand. The Armenia earthquake in which approximately 25,000 lives were lost, made headlines for several days. The South Pacific earthquake of 23 May 1986—almost 25 times greater—was lucky to get a 2 inch article on page 12.

To address the different aspects of earthquakes, several attempts have been made to assign scale values representing the physical destruction. The Modified Mercalli Scale is based on damage sustained and represents one attempt to compare earthquake intensities. The scale has 12 categories, ranging from being felt only under special conditions, to visible ground waves and objects tossed into the air. Assigning scale values becomes a judgment call. When man-made features are not present, much of the plane of reference is missing. Two sections of the same large city may even be assigned different earthquake intensities as a result of the soil or rock foundation supporting the structures. Since physical damage is the observed signature, scale values normally decrease away from the epicenter. The Mexico City earthquake is an exception where the rupture occurred some 200 miles to the west. Some communities along the Pacific coast, much closer to the earthquake center, survived with little damage.

San Francisco 1906 Earthquake. On 18 April 1906, an 8.25 earthquake hit the San Francisco area when the friction lock holding a segment of the San Andreas Fault failed. The Pacific Plate lunged to the Northwest. Ruptured electric and gas lines triggered destructive fires in the city. The estimated direct damage reached $20 million with another $400 million in fire damage. The death count reached 315 with another 352 unaccounted. The same earthquake today would have a greater price tag, even with all the steps that have been taken in anticipation of another "Big One."

Billions of seconds of evolutionary creeping between the Pacific and North American tectonic plates is occasionally interrupted by a thrust of several feet over a period of a few seconds. Stress accumulates until the friction lock can no longer resist. In the same way a heavy box alternately hesitates and slides when being pushed or pulled across the floor. Earthquakes demonstrate how both evolutionary change and cataclysmic surges play a part in molding our earth.

Loma Prieta 1989 Earthquake (San Francisco). At 6:04 P.M. on 17 October 1989, an earthquake hit the San Francisco area. The epicenter was about ten miles north of Santa Cruz near Loma Prieta. For fifteen long seconds, the earth in the San Francisco area shook violently to the tune of a 7.1 magnitude earthquake (originally rated 6.9).

Sports fans, tuned in to watch the World Series game between San Francisco and Oakland, got a live shot of the vibrations and the reaction of the crowd. As the evening progressed, television viewers had a ringside seat to witness scenes from the collapsed I-880 Nimitz Freeway double deck overpass, the collapsed section of Oakland Bay Bridge, and the fires in the Marina District. Both the Marina District and the collapsed freeway were on soft landfill, creating the additional instability and destruction.

Early estimates of 271 killed and 1,400 injured were later revised to 63 killed (39 on the Nimitz Freeway) and 3,286 treated in hospitals. The physical damage estimate was set at

$7.1 billion with 13,892 homes lost. An aftershock of 5.2 hit just 40 minutes after the initial earthquake. Over the next ten days, over 4,000 aftershocks were recorded—20 over 4.0.

Scientists searched the area of the identified epicenter and found a 1,000 foot tear in the ground. Actually, there were several parallel scars about 1,225 feet apart. Scientists were quick to point out that the 17 October earthquake was not the "Big One" they were anticipating. The rupture turned out to be on the Zayante Fault, a subsidiary of the San Andreas Fault system. Contrary to early speculation, the surface scars did not confirm a northward movement of the Pacific Plate from the North American Plate. Only time will tell, if the rupture adequately relieved the stress built-up between the two tectonic plates to delay the "Big One". Will the stress simply to shift to another segment of the San Andreas Fault system and accelerate its slippage time table? The lack of evidence of relative north-south slippage would suggest that most of the stress built-up remains intact.

Los Angeles area earthquakes. Unlike the 1906 San Francisco earthquake that allowed the Pacific Plate to slide north along the San Andreas Fault, the 5.9 magnitude Whittier Narrows earthquake of 1 October 1987 was a dip slip fault deep within the crust. The slippage, almost perpendicular to the San Andreas Fault, reflects the compressive component of stress between the Pacific and North American plates. The damage reached $358 million with 8 dead.

Other major earthquakes in the Los Angeles area include Santa Barbara (6.2 magnitude) in 1925, Long Beach (6.3) in 1933, and San Fernando (6.6) in 1971. The San Fernando earthquake may be best remembered by the pictures of 5 highway overpasses that crumbled. Its price tag reached $500 million with 64 lives lost.

The Northridge (6.8 magnitude) earthquake of January 17, 1994, caught most Los Angeles residents asleep. It was on a deep fault some 12-20 kilometers below the surface. As with other earthquakes in the Los Angeles metropolitan area, the Northridge slippage was on a dip slip fault, as opposed to the strike slip fault of the San Andreas that passes east of the city. Its primarily vertical shock waves provided a ground motion estimated at 40 percent greater than normal for earthquakes of its magnitude. Television cameras provided a graphic account of the physical destruction during the hours following the earthquake. We can only speculate on the loss of life, had the earthquake hit during the rush hour.

An earthquake, equivalent to the San Francisco earthquake of 1906, hit 60 miles north of Los Angeles near Fort Tejon in 1857. Over a 3 minute period, 217 miles of the San Andreas fault in south-central California ruptured. Only two people died from the earthquake—far less than if the same magnitude earthquake occurred today.

Looking back, Californians still talk about the 1906 San Francisco earthquake as the largest in recorded history (Fort Tejon did not get the press). Looking ahead, the question remains; where, how, and when will the "Big One" hit?

California's strongest quake to hit in the last 40 years struck near Landers California, at 4:58 a.m. on 28 June 1992 at a magnitude of 7.4. Being several miles east of Los Angeles, California, the city was spared another costly disaster, even though the local communities in the desert were hit hard.

Anchorage 1964 Good Friday Earthquake. Prince William Sound, Alaska was hit with an 8.3 magnitude earthquake on 27 March 1964 (Good Friday). Undulating waves shook Anchorage and opened 30 foot fissures. One wing of a school dropped 20 feet into a depression. Stores on one side of Fourth Avenue, in the main business district, dropped 11 feet destroying the buildings while leaving the other side of the street nearly intact. The Governor pleaded for $500 million to help with the rebuilding. It is remarkable that only 33 persons were confirmed killed with an additional 82 presumed dead.

Mexico City 1985 Earthquake. Even more destructive was the earthquake that shook Mexico City at 7:17 A.M. on 19 September 1985. The earthquake, registering 8.1 on the Richter scale, shook the ground for about one minute. Some 36 hours later a 7.5 magnitude aftershock followed.

Mexico City was over 200 miles from the epicenter of the west coast of Central America. Shock waves from the rupture were magnified by the soft sediments on which much of Mexico City is built. The final toll was over 8,000 killed or missing, 30,000 injured, and 50,000 left homeless out of a population of over 18 million. Property damage to 3,300 buildings was estimated at over 4 billion dollars.

The Mexico City earthquake did not come as a complete surprise. The area has been seismically active in the past. Some 42 earthquakes of magnitude 7.0 and above struck the Mexico segment of the San Andreas Fault this century. This compares with only 7 earthquakes of magnitude 7.0 and above in California.

New Madrid 1812 Earthquakes. The strongest earthquakes in United States' recorded history were not associated with the relative movements between the North American and Pacific plates. They were in the Mississippi Valley. Between 23 January 1812 and 12 February 1812, three major earthquakes measuring 8.6, 8.4, and 8.7 on the Richter scale rocked the New Madrid, Missouri area. The channel of the Mississippi was dramatically altered. The area is still seismically active with an average of one minor shock every 48 hours. A repeat of the earthquake sequence of the 1812's magnitude today would cause far more physical damage as population and industry have increased many fold.

The 17 October 1989 Loma Prieta earthquake has prompted a renewal of earthquake preparation activity along the New Madrid fault zone, as well as in other earthquake prone regions. The natural uneasiness of the area residents made them the victim of self-appointed experts and psychics that gambled on gaining notoriety by predicting the crust rupture time. Aided by much free publicity by the media, unnecessary fear was generated and money expended because of a prediction of a major earthquake on or about 4 December 1990. This earthquake never occurred.

Middle East Earthquakes. On 7 December 1988, an earthquake struck Armenia (in the Caucasus Region of Russia) causing $16.2 billion in damages. It destroyed some 58 villages and towns. The death toll reached 25,000 with an additional 19,000 injured. From the above statistics, it appears that the earthquake intensity should have been much greater than its 6.9 rating (compare with the 7.1 Loma Prieta earthquake). The type and quality of construction, and the region's geology are blamed for the high loss of life and property damage. The Armenian earthquake was a thrust fault caused by the North driven Indo-Australian plate lifting the Asian Plate about 6 feet. Several other destructive earthquakes have hit the seismically active Middle East since the Armenian earthquake.

Omissions

Any discussion of earthquakes is subject to criticism of omissions. According to loss of life, China's earthquake of 1546 leads the list with approximately 80,000 killed. Four major earthquakes in China between 1920-1976 account for 1,100,000 killed. Japan lost about 450,000 lives in 3 destructive earthquakes between 1803-1923. India's "Big One" took 300,000 lives in 1737. The Mid-East, Mediterranean, West Indies, and South America (Andes Region) have had several earthquakes with tolls over 20,000 each.

Of course, lives lost is often a misleading measure of the intensity of earthquakes. The loss of life is more related to the population density, geologic conditions, building construction, and earthquake-driven tsunamis. However, to the family that suffered the only person killed in a minor earthquake, the event is still cataclysmic.

Human-induced earthquakes

The amount of stress that builds up along an existing fault surface is a function of the friction between the crustal blocks. This point was well illustrated by David M. Evans, a consulting geologists. He correlated an increase of minor earthquakes in the area northeast of Denver, Colorado to fluids that were pumped deep into the ground under pressure at the Rocky Mountain Arsenal. Mr. Evans reasoned that the fluids, under pressure, were serving as a lubricant that facilitated the fault movements. After many long debates, Evans was able to persuade authorities to stop the pumping on a trial basis to check his theory. When pumping ceased, the frequency of earthquakes dropped. This provided the first concrete evidence in support of his theory. When the pumping was resumed, the frequency of the earthquakes also accelerated. An additional test of the theory was conducted in the Rangely, Colorado area, with comparable results. Both scientists and government officials accepted the theory, based on the demonstrated evidence, and permanently concluded the well pumping procedure.

In the long run, discontinued pumping might result in a much larger future earthquake for the Denver area, if the forces responsible for the stress continue to build. Although the fault, lubricated by fluid injections, apparently allowed the crustal blocks to slip, no slippage would have occurred, if the stresses between the blocks were absent. Now, without the fluid lubricant, a higher stress level will have to build before it is released as an earthquake.

The idea of lubricating known faults to decrease the degree of shock has been suggested as a means of reducing the magnitude and increasing the frequency of earthquakes. In theory, an earthquake measuring 8.0 on the Richter scale could be replaced by 30 earthquakes of 7.0 magnitude, or 900 earthquakes of 6.0 magnitude, or 27,000 earthquakes of 5.0 magnitude, etc. To replace an 8.0 magnitude earthquake every 100 years with 5.0 earthquakes about every 2 weeks does not present itself as an attractive option. Another problem facing scientists, specially in areas such as San Francisco or Los Angeles, is that the fault zone lubrication could release existing stresses to trigger an earthquake of 8.0 or larger magnitude.

Although planned lubrication of known faults could still prevent an even greater future earthquake, it would be political suicide for the authorities to implement such a policy. The legal ramifications would flood courts for decades to come. If the idea of pumping water under pressure into a known fault zone is ever to be adopted, it should be done immediately after a natural, large earthquake has released most or all of the existing stresses. As the future stresses build up, they would be released in a series of smaller earthquakes.

The Hollister, California area, south of San Jose and midway along the San Andreas Fault, could be a possible site for such tests. Near Hollister, the San Andreas Fault has been slipping in small segments. Fault creep prevents the build-up of stress needed for a major earthquake. Any artificial earthquake should, therefore, be relatively small and cause limited damage to the sparsely populated area. In theory, the frequent but reduced magnitude earthquakes resulting from lubricating the Hollister segment of the San Andreas Fault would hardly be felt. Still, would you care to be the government official that approved them?

Many additional factors must be addressed, if lubricating a faults is to be considered. The incremental release of stress along the different segments of the San Andreas Fault transfers some stress to adjacent segments of the fault zone. Justification for inducing fluids to reduce the earthquake dangers in one segment, must be weighed against any increased danger imposed on adjacent areas. Lives and property damage at stake are primary considerations. Even then, no individual or group of individuals can take a chance that favors the lives of one community over another.

Predicting Earthquakes. A priority of earth scientists is to Increase our knowledge of warning signs for pending volcanic eruptions and major earthquakes. There are hopes of reducing the potential for human tragedies associated with the cataclysmic events. Scientists, however, need to be concerned with how they will employ new signs of pending earthquakes. It is very important that any official predictions of earthquakes or volcanic eruptions be based on very solid scientific evidence. Nothing could set back an early warning program more than to issue warnings that never develop, thus generating the impression that the scientists are just crying wolf.

The only formally endorsed earthquake prediction so far attempted, is a warning of a pending earthquake in the 5.0+ range for the Parkfield, California area. The Parkfield prediction is based on a pattern of earthquakes in the area occurring about every 22 years. The latest earthquake of the series was a 5.6 rupture in 1966.

Late in 1993, scientists saw signs of the overdue eruption. The signs soon subsided, and the scientists were back on hold waiting for another sign for this overdue earthquake.

There are many activities associated with implementing an earthquake prediction program for an urban area, such as San Francisco, Los Angeles, or St. Louis. They include extensive evacuation of structures not designed to resist earthquake vibrations, closing of highways, bridges, and tunnels along the fault, shutting off utilities to reduce the chance of fires, and setting up emergency facilities to handle injured and displaced people. The expenses and natural panic associated with such an evacuation plan, the hardships imposed by power losses, and the problems of looting associated with many disasters, dictate that warnings must be close to 100 per cent certain—both in time and magnitude—before they can be issued. The test prediction at Parkfield does not involve such extensive precautions, consequently, it is considered a safe test area. It is a good thing that the precautions did not involve elaborate measures, since it did not develop as the scientists anticipated. A stepped grading system is in place to reflect the urgency of each warning.

Animal behavior has been suggested as a forecaster's clue for pending earthquakes. However, it is subjective to exactly recollect animal behavior after the fact. The chore for scientists is to identify the animal responses that only occur before an earthquake. Even if an animal's behavior can be attributed to pending earthquakes, it may not provide sufficient time to warn the population.

One signal that emerges as a possible sign of pending earthquakes is variation in electrical charges. Greek scientists claimed successful measurements for predicting earthquakes. Theodore Madden, of Massachusetts Institute of Technology, reported detecting electrical charges in the ground before the Whittier Narrows earthquake in Los Angeles. Anthony Fraser-Smith, an electrical engineer at Stanford, reported that an ultra low frequency radio receiver located near Corralitos (just 4 miles from the Loma Prieta earthquake) detected unusually strong electromagnetic signals as early as 5 October 1989. On the day of the Loma Prieta earthquake, a sharp rise of intensity was observed at 2 P.M., followed by a brief fall. Just before the earthquake the electromagnetometer wave intensity skyrocketed.

The mechanics of earthquakes are discussed in Appendix D for force vectors acting on the fault surface and friction locks.

Earthquakes Help Foretell Pending Volcanic Activity

Earthquakes are one of the early signs of pending volcanic eruptions. When the potential of an eruption is known, scientists monitor the seismic activity deep within the volcano. As magma pushes up into the crust, existing fractures expand and new fractures open. Each rupture sends out earthquake vibrations. A flurry of earthquake activity, deep under a volcano, signals magma movement and the potential for an eruption.

For our purposes, we will limit our discussion of specific volcanoes to the few noted in chapter 3 and the brief comments below. Just as selecting earthquakes for our discussion, the volcanoes missing from this list would fill volumes. Who could omit Nevado del Ruiz in Armeno, Columbia where 22,000 lost their lives in 1985; or Pelee in Pierrez, Martinique (death toll 28,000); and La Soufriere in St Vincent, Martinique (death toll 15,000) in 1902. In general, volcanoes occur in many of the same areas as the largest earthquakes. The death toll for major volcanoes, however, appears to be about one order of magnitude lower than for the cataclysmic earthquakes.

Pinatubo volcano in the Philippines provided several warning signals before its series of eruptions began. Pre-eruption precautions undoubtedly saved many lives. The heavy rains soaked the unconsolidated ash deposits and triggered mud slides.

Mount St. Helens. The eruption of Mount St. Helens [Indian named her Loo-Wit (lovely maiden) and Tah-one-lat-clah (fire mountain)] on 18 May 1980, came after many years of dormancy. With over 20 eruptions in the past 4,500 years, it did not come as a big surprise to the scientists who reviewed the seismic records and other signs of the pending eruption. The exact time and magnitude of the eruption were beyond the scientists' ability to predict. Some 61 lives were lost, primarily because the eruption was greater than anticipated. Of course, stubbornness was the biggest factor for the death of Harry R. Truman, a long time resident of Spirit Lake. He had received sufficiently warning of the pending disaster.

Mount St. Helens' eruption was the first in the United States since Lassen Peak's eruptions between 1914-1917. Mount Rainier erupted sometime between 1820-1854. At the first sign of activity in any of the Cascade volcanic mountains, scientists will move in with their sensitive instruments to closely monitor changes.

Long Valley, California, east of Yosemite National Park, was hit with four 6.0+ magnitude earthquakes during a three day period in May 1980. This was followed by thirteen 5.0+ earthquakes from May 1980 to November 1984. The ground in the area raised an average of about 35 millimeters per year from 1982-1985. An increase in seismic activity in 1989 and 1990 intensified the scientific attention. They are now watching for a swarm of distinctive earthquakes similar to the type observed preceding recent volcanic activities in Japan and Alaska. The last volcanic eruption in Long Valley occurred about 600 years ago. Because of its volcanic history and the most recent signals, scientists are closely monitoring the area as reported in Richard A. Kerr's article, "Long Valley is Quiet but Still Bulging."

Palmdale bulge. An area north of Los Angeles centered near Palmdale, California (referred to as the Palmdale bulge) rose about 25 centimeters during 1960-1962 and then bulged up another 15 centimeters during 1972-1974. It is too early to make any solid predictions on the potential of volcanic activity. The size and shape of the area fit the scenario for magma intrusion forming a batholith or laccolith, if the bulge continues to grow.

Yellowstone Caldera. The Yellowstone Caldera rose an average of 15-20 millimeters per year since 1923. The bulging of the Yellowstone Caldera provides evidence that its resurgence is still taking place. If pressure continues to increase over the next few thousand years, another giant eruption could occur. Other options include occasional eruptions of smaller explosive volcanoes, the eruption of flood lava, the formation of a batholith or laccolith, or simply a reverse of the present trend to cool down and become extinct.

Although scientists are gaining new knowledge every day, to help in forecasting earthquakes and volcanic eruptions, the nature of the final rupture makes it impossible to predict the exact time and intensity. Modern history is replete with reports of cataclysmic events catching many people off guard, simply because we cannot predict the time within hours, days, months, or even years.

A Case for Axis Surges

In chapter 4, we used a hypothetical one-degree axis shift to illustrate the effect of geoid changes. Using the present drift rate, it would take about 1,250,000 years for the axis to drift one-degree in a straight line. Who is to say that weight shifts will not cause the pole to meander instead of drift along a straight line? Why should axis surges not greatly reduce the 1,250,000 years required for one-degree axis movements. It is too difficult to project axis movements over a million years on the strength of less than 100 years of observed data. It would be a similar endeavor to program a car's steering for all the proper turns along 200 miles of highway, before safely pulling into your driveway at the end of the route.

Both earthquakes and volcanoes represent the sudden release of accumulated stresses in the earth's crust. In the case of a segment of the San Andreas Fault, strain gages detect a very minor creep in fractions of an inch that suggest to scientists the stress build up. At the time of rupture, the fault slippage can be measured in feet.

Using the earthquake analogy, the axis wobble is the strain gage that indicates the accumulation of eccentric weights on our planet. As weight shifts increase the dynamic imbalance, the amplitude of the wobble increases. The greater the wobble amplitude, the greater the amplitude of the magma waves that massage the crust. Occasional axis surges equate to the release of stresses accumulating on the earth's axis from dynamic imbalance.

For axis wobble, the most rigid regions (thick continental crust) of the tectonic plates will respond to the magma wave moving up and down as a block with little or no internal flexing. The thinner crust will have a greater tendency to flex over large areas—much like the deck of a bridge bowing slightly as the train passes.

Along lines of relative crustal weakness (fault zones), such as the plate edges, the plates will flex like a hinge. After thousands of years of flexing along a crustal hinge, fatigue will eventually set the stage for a major rupture. The North American Plate has at least three natural hinge lines internally within the plate. They are along the west side of the Wasatch Range in Utah and Idaho, along the New Madrid Fault zone in the Mississippi Valley, and a hinge line just east of the continental shelf off the east coast of North America, where the thinner ocean crust joins the thicker continental land mass. The 31 August 1886 Charleston, South Carolina earthquake that killed 110 people provides evidence of such crustal weakness along the continent's edge.

If the flexing hinge line is also near the shift meridian (as indicated by long term axis drift), the release of positive magma pressure could create a new mountain range, batholith, laccolith, flood lava, a string of volcanic eruptions, or some combination of the above. In the negative quadrants, the compressive forces will either buckle the crust into a folded mountain range, e.g., in the Appalachians, collapse it to create or extend an ocean trench, or take up the excess crust by forming an overthrust belt.

We can only speculate on how fast an axis surge will move and how long it will take for the magma to respond to an axis surge. We do know that earthquake and volcano pre-rupture stress accumulates over a long period. We also know that some post rupture stress in the crust remains. If all stresses were released, there would never be aftershocks. For our axis surge scenario, the ocean waters will adjust to the new geoid in a matter of hours or days with massive tsunamis sloshing back and forth across the ocean basins. The high viscosity of the magma, however, will require a much longer period to stabilize the magma seas—possibly measured in decades or centuries.

The response of ocean waters, magma seas, and crustal deformation to an axis surge will constitute an additional massive weight shift in the earth's dynamic balance occurring over a short time. Weight shifts from ocean and magma sea adjustments will trigger even more axis shifts and magma pressure variations. Once the axis shift starts, its movement builds on itself until some degree of stability is eventually achieved. For earthquakes, a series of aftershocks usually precede stability.

Tectonic plates will have their greatest degree of freedom, or least relative stress, immediately after an axis surge. The increased crustal stress between plate edges that are near the shift meridian is relieved by the rupturing. The tectonic plate edges near the neutral meridional and equatorial planes carry less accumulated stress. Consequently, they will experience little or no shift. Without stress to re-rupture the plate edges, fracture zones have more time to heal. The intrusion of magma into crustal fractures, to form dikes and sills, also serves as a Nature's body shop by welding some of the broken parts.

The Ural Mountains in the Russia, the Appalachian and Rocky Mountains in the United States, and the Trans-Antarctica Range near the South Pole, now far from an active plate edge, have accomplished a high degree of healing since their more active days. The theory of plate tectonics explains mountain building primarily as a product of plate collisions. The Himalayas, Alps, Cascades, and Andes are associated with active plate edges. Plate tectonics assumes mountain building is a slow evolutionary process. The Dynamic Axis Theory incorporates the plate tectonic concepts, including collision mechanics, but extends it to incorporate the occasional surges and cataclysmic evolution. The Dynamic Axis Theory also provides for the interchange of crustal forces between tension and compression to create new active mountain ranges and allow old ranges to heal.

.

7   EVIDENCE OF CATACLYSMIC AXIS SHIFTS

It should not be difficult to find some concrete evidence, if the axis shifts in surges. We do not have to search far. Marine terraces, guyots, ice ages, mass extinctions, frozen mammoths, sedimentary beds, submarine canyons, fjords, and magnetic reversals each provide clues that support the concept of cataclysmic change. A single scenario, based on an axis surge, can explain all the above mentioned geological events.

Marine Terraces

Ancestral ocean beaches found above and below the present sea level indicate that either the geoid has changed dramatically several times over past geologic ages, or large regions of crust have had rapid vertical movements. From chapter 2 we can see that the Dynamic Axis Theory provides a simple answer to an ever-changing sea level and land-ocean boundaries, even without the contribution provided by crustal subsidence or uplift. In chapter 3 we discussed vertical movements in the crust in response to variations in the magma sea pressures. Crustal gymnastics will always be a part of the equation, although the easiest way to explain multilevel marine terraces is to rotate the geoid as we did for the hypothetical 1-degree axis shift.

Peru-Chili marine terrace

Charles Darwin's Journal contains information about a marine terrace along the west coast of South America. Near Lima, Peru the terrace is about 85 feet above sea level. Near Valparaiso, Chili the same terrace is about 1,300 feet above sea level. The computed values for our hypothetical 1-degree axis shift revealed a sea level 478 feet higher for Lima, and 1,121 feet higher for Valparaiso. Both show the terraces plane tilting up toward the South. The computed change in geoid differs from the actual marine terrace (Lima 353 feet higher and Valparaiso 179 feet lower). This can be explained by differential crustal response to magma seas.

New Zealand marine terraces

Marine terraces in New Zealand, just north of 45 degrees south latitude, are also tilted with their higher elevations located in the South. This is consistent with Table B-1 (Appendix B) that shows an axis shift increasing the geoid elevation in the positive quadrants from the equator to 45 degrees latitude. A corresponding decrease in elevation occurs between 45 degrees latitude and the pole. In the negative quadrant, the geoid response is reversed.

Evolutionary change in sea level would destroy terraces

Our hypothetical example of sea level changes in chapter 4 was based on a one-degree axis shift. This shift required 1,250,000 years based on the present axis drift rate. The possibility of an occasional axis shift surge was not considered. The marine terrace evidence does not support, however, the assumption of a gradual axis drift. Terraces represent periods of relative stability in sea level over an extended period, while wave action carves the beach terrace. Elevated and drowned terraces, therefore, indicate a sudden and dramatic geoid change: one shift to bring the geoid to the terrace level, another to move it away. A gradual change in sea level would have obliterated the individually identifiably terraces. The forces and mechanics of crustal gymnastics needed to account for marine terraces (without a cataclysmic change) require an even bigger imagination than mine.

Ice Age marine terraces

Fluctuations in ocean waters locked up in ice caps can only explain terraces that are nearly parallel to the present sea level. Ocean waters locked up in the North American and European ice caps can only account for marine terraces that are within a few hundred feet below the present sea level (even if we ignore any isostatic rebound of the unloaded ocean crust). The flip side of this discussion is flooding that would occur along all coasts, if the ice caps of Antarctica and Greenland melt. Although variations in the volume of waters locked in glacial ice will affect sea level, isolating the arguments for terrace formations around such a relatively minor change in geologic history ignores the need to explain evidence of geoid changes that are measured in thousands of feet.

Terrace scars cut by a lake rather than by ocean waves are good examples. The terraces cut by Lake Bonneville (the Great Salt Lake Basin) represent water level changes over a relatively short geologic age. During the Ice Age, the region was much wetter and Lake Bonneville was full to the brim. Lake Bonneville overflow waters drained out to the north into the Snake River. Evidently, a period of rapid erosion cut the drainage channel deeper and lowered the lake surface to the Provo terrace level. Both the Bonneville and Provo terraces can easily be seen along the base of the Wasatch Mountains from the Salt Lake City and Provo area. Continued evaporation dropped the lake surface to its present level. It fluctuated a few inches or feet each year, depending on the amount of precipitation in the basin. No prominent lake shore scars (terraces) equivalent to the Bonneville and Provo levels were left as the evaporation gradually lowered the lake surface to its present level. This demonstrates the evaporation was relatively steady and rapid.

Neither axis drift or tectonic plates, propelled at a snail's pace, can account for the formation of marine terraces found well above or below the present sea level. Marine terraces can only be formed by relatively sudden change in sea level followed by extended periods of little or no change. Axis surges, as presented in the Dynamic Axis Theory, provide a simple explanation for large changes in the geoid over short periods of time.

Hot spots, seamounts, guyots, and volcanic islands

Guyots are flat-topped seamounts that are formed by erosion and coral reef growth at or near sea level, but now drowned. Many are under hundreds of feet of ocean waters. The supporting structures for the guyots are volcanic cones resting on the ocean floor. Some of the larger seamounts have grown to extend above sea level and form islands. Only when the ocean depths over the volcanic cones are very shallow, or the cone projects above sea level, can the coral reefs grow by attaching to seamounts.

Enrico Bonatti and Kathleen Cranc's article "Ocean fracture Zones" presents coral evidence of dramatic changes in sea level. They noted a sliver of crust now 1,800 feet below sea level on the south side of the Vema Transform Fault, at 11 degrees north latitude in the Atlantic Ocean. It is capped by reef limestone similar to those ringing the Bahamas Islands. At the eastern intersection of the Romanche Transform Fault with the Mid-Atlantic Ridge they found another summit 3,000 feet below sea level that is capped by a fossil reef. Both could be explained by a sudden drowning. Gradual drowning by a rising sea level (or subsiding land) would have allowed coral to continually build on the sinking coral base until the guyot's size, water temperature, or some other physical change ended their growth.

Extending north-northwest from the Hawaii Islands is a row of seamounts that led scientists to the hypothesis that the crust moves over a fixed hot spot in the asthenosphere. Inspection of the National Geographic Society's Physical Map of the Pacific Ocean reveals the existence of several aligned chains of seamounts. The Hawaiian Ridge from Midway Islands to Hawaii is the most prominent. The hot spot is presently under Hawaii, and it is responsible for the volcanic activity, including a new island in the making that is still well below sea level. The Emperor Seamount Chain appears to be an extension of the Hawaiian Ridge but with a dog leg giving it a more northern alignment. It has been suggested that the two chains were formed by the Pacific Plate passing over the same hot spot. A change in the direction of Pacific Plate movement is credited with creating the angle in their alignment.

Several minor lines of seamounts exist throughout the Pacific Basin, but with random alignments. About halfway between Hawaii and Midway Islands, the Necker Ridge intercepts the Hawaiian Ridge at an angle of about 60 degrees. To the north of the Hawaiian Ridge the Musicians Seamounts form a random pattern. To the south and east of the Hawaiian Ridge are several short strings of seamounts and islands, and many that are randomly positioned. In the Southern Pacific, a string of seamounts appears to be a westward extension of the Eltanin Fracture Zone. (Aligned seamounts are often associated with transform fault zones.) Other strings of seamounts are aligned parallel to the fracture zones associated with the East Pacific Ridge, but without direct evidence of fractures. This could simply be a lack of detail in the ocean floor map. The Atlantic and Indian Oceans have some randomly positioned seamounts. There are, however, no major seamount chains that are comparable to those found on the floor of the Pacific Ocean.

If we assume the Pacific Plate is moving northwest over more than one fixed hot plume (the Hawaiian Ridge being one), the alignment of seamounts and volcanic islands created by each hot spot should be near parallel. Of course, short-lived hot spots could account for isolated volcanic cones but not a ridge alignment like the Necker Ridge that forms an angle to the Hawaiian Ridge. If there was any rotation in the plate's motion as it traveled northwest, it would have curved the alignment of each string of volcanic islands and seamounts. They would, however, remain basically parallel and concentric about the plate's center of rotation. The theory that the same hot spot formed the Emperor and Hawaiian chains is inconsistent the possibility that each short oblique seamount chain was formed during different geologic epochs, when the Pacific Plate was traveling a different path.

As mentioned earlier, the magma currents should respond much like a slow motion atmosphere with the hot spot representing magma's tornadoes. Magma tornadoes could account for non-parallel seamount alignment, and short lived hot spots that produce randomly positioned seamounts. High density seamounts would represent a zone of violent magma activity where cold and hot magma currents collide—the equivalent of tornado alley. The idea of magma tornadoes is consistent with Clement Chase's work mentioned earlier.

Non-parallel linear chains and randomly distributed patterns of seamounts make it difficult to accept the notion that they are created by the tectonic plates moving over stationary hot spots.

Similar to marine terraces, guyots and submerged reefs signify an extended period of near constant sea level. The lack of coral reefs higher and lower on the volcanic pedestals suggests that the change in sea level was not a slow evolutionary process.

Climate and Continental Glaciers

Geologists know that about 18,000 years ago, the ice caps that covered much of northern North America and western Europe were near their peak. The North American ice sheet retreated for one of many interglacial periods before making another advance about 12,000 years ago. Estimated dates for the start of the meltdown, as well as the length of time it took to melt the ice sheet vary considerably, depending on which text is consulted. Recent discoveries have even challenged the accuracy of past carbon-14 dating results. Some carbon-14 derived dates may need to be adjusted backwards several years. "Geologic Time", a United States Geological Survey brochure, pegged the age of a forest bed near Milwaukee, Wisconsin at 11,640 years. This correlates with one of the last advances of the continental ice sheet.

Estimates for the meltdown vary from as short as 2,000 years to 8,000 years or more. This probably reflects the site selections where the clues were observed. The meltdown period, regardless of the estimate, represents just a blink of the eye in geologic time. For our purposes, we are only interested in the fact that the meltdown period was short.

It took a relatively sudden and dramatic change in climate to trigger the meltdown of the two mile thick ice sheet. The icy landscape provided refrigeration that kept any slow moving air above it in the deep freeze. Radiant heat from the sun by itself would have been reflected off the white landscape. Similarly, air temperatures stay well below freezing on a bright winter day with snow cover on the ground and no wind. Temperature inversions that lock the cold air in valleys and natural pockets reveal just how little warming is contributed by the sun alone.

For the ice sheet to start its decay required some combination of higher temperatures during the summer months, a longer summer, less snow during the winter, ice-eating chinook winds, and warm summer rains on the ice sheet. The ice sheet at the start of the period had to melt, as well as the winter snows that fell during the meltdown period. Warmer atmospheric currents had to supply external heat, specially during the early stages of the meltdown. One possibility is an axis surge that altered the land-ocean boundaries, and lowered the latitudes and the elevations. This changed the prevailing air current patterns, temperatures, and moisture content. If the change was rapid, the meltdown could be accomplished with about half the increase in air temperature. An average 10 degree warming over the meltdown period would generate about the same ice cap decay as a gradual 20 degree increase in air temperatures over the same period. Lower moisture, specially during the winter months, would also contribute to a rapid meltdown. We can apply this "what if" game to the trigger that produced the ice sheet in the first place. An axis surge reversal is all we need to provide higher latitudes and elevations, with prevailing air currents bringing in moist air to fall as snow. Even if the idea of an axis surge could only explain the formation and decay of ice sheets, it would be worthy of serious consideration by earth scientists. The fact that the Dynamic Axis Theory seems to provide answers for many perplexing earth science questions gives it much additional strength.

Every now and then we run across a statement that, at first glance, seems contrary to logic. That is usually because we have not had the occasion to fully think about it. A recent report about the drilling of cores in the Greenland and Antarctica Ice sheets noted that snow samples from deep within the ice sheet were deposited 14,000 years ago. If earth was in an Ice Age, would the Greenland Ice Cap not have been even thicker 14,000 years ago? Is the present ice sheet the remnant of the Ice Age? Why is the snow deposited near the end of the ice age so deep in the present ice sheet? Are the ice sheets still growing during this interglacial period?

The answer, of course, is that glaciers melt from the bottom, even during their growing stage. Temperatures just below the upper surface of the Antarctica Ice Sheet were measured at about -20 degrees Fahrenheit. That is considerably warmer than the winter atmospheric temperatures above the ice. Eskimos took advantage of the same insulating characteristics of ice when they built temporary igloo shelters. Lost skiers seek cover in snow caves. Temperatures at the base of the ice sheet, insulated even more from the atmosphere and warmed by heat conducted up through the crust from the magma interior, were near 32 degrees Fahrenheit.

Glaciers grow as the compacted snow turns to ice and more snow is deposited on top. The weight of the glacier moves it down hill like an ice river. Friction of the moving ice generates some heat. Melting at the base of the glacier facilitates the movement of this river of ice. This suggests we look for the oldest ice at the bottom of the glacier near its delta.

The North American ice sheet did not grow and decay over a geologic day. Several interglacial periods occurred that melted much of the glacier only to be reformed again. If a continental ice sheet can form or melt in less than 20,000 years, we must look at geologic records of past ages in a completely new light. Geologists consider even the end of the Cretaceous period 65 million years ago as almost modern history. During a one million year period, some 65 million years ago, several ice sheets could have appeared and disappeared. Ice sheets in different parts of the world could have occupied lands that have long since been eroded away or buried under new sediments, destroying or hiding evidence of their existence.

The problem for geologists is evident when we realize that their clues come primarily from the crustal surfaces that just happen to be exposed today. Thanks to erosion that has exposed the multi-layers of crustal formations, such as in the Grand Canyon of the Colorado, considerable insight has been gained about geologic history. Interpolations between detailed geologic studies are used to fill the gaps. Well core logs, mine shafts, and tunnels have only slightly expanded the knowledge of general geologic history. Well logs, for example, are limited in their contribution to the detailed local geologic history. A well log could easily miss the significance of the petrified forests, dinosaur burial grounds, or evidence of glacial activity.

A sudden and dramatic change in climate about 12,000 years ago was necessary to trigger the rapid meltdown of the ice sheets over northern North America and western Europe. A surge of the earth's axis, as conceived by the Dynamic Axis Theory, can explain the climate change needed as latitude, elevation, and the land-ocean distribution that determine regional climate conditions.

That the planet recovered from an Ice Age in much less than 10,000 years raises questions on how many glacial periods are probably going undetected in geologic history spanning billions of years.

For the ice sheet to build, the only condition to be satisfied is that more snow falls during the winter months than melts during the winter. Heavy snow is not essential as we can see in Antarctica where precipitation is much less than for many non-glaciated areas. Cold weather by itself will not generate an ice sheet. If it did, there would be more ice covered lands in the Arctic Region. The Brooks Range areas in northern Alaska and Siberia are good examples of desert arctic with its very light precipitation. Alaska's major valley glaciers are found at lower latitudes in southeastern Alaska. Here moisture laden clouds are cooled by elevated land triggering heavy snows and preserving part of the winter snow from the summer melt.

Under normal conditions, the rate of glacier decay as a function of temperature will be much faster than the buildup. A decrease in the average temperatures of 20 degrees during the winter months will do little to accelerate the buildup of a glacier. A 20 degree increase in average temperatures over the summer months will markedly affect the decay rate. Ice sheet growth could, however, be relatively fast, if very heavy winter snow falls combine with heavy cloud cover and Arctic cooling breezes that dramatically reduce summer melting.

Mass Extinctions

David M. Raup of the geophysical sciences department at the University of Chicago, published an article in Science, entitled "Biological Extinction in Earth History." He postulated that virtually all plant and animal species that ever existed on the earth are extinct. Many species simply fall by the wayside for one reason or another. Of the greatest interest, however, are the major periods of mass extinction that dot geologic time. Dale A. Russell's article "The Mass Extinctions of the Late Mesozoic" reminds us that mass extinctions are a recurring event over geologic history. Attempts have been made to see if a uniform cyclic pattern is responsible, but solid evidence is elusive. Five large mass extinction events have occurred over the past 600 million years. The Cretaceous-Tertiary (often referred to as K-T) mass extinction is the best documented one. Dinosaurs represented only a small part of that extinction event, but they have attracted the most attention.

Mass extinction events often take their toll on the larger terrestrial animals. Tropical biotas, discussed below, are more prone to extinction. In the search for a cause of mass extinction periods, environment and sea level changes are frequently mentioned. However, no causes for the climate change are offered. The axis surge events of the Dynamic Axis Theory used to explain geoid changes and crustal gymnastics would trigger dramatic climate change. They set in motion a series of interrelated events that contribute to selected species extinction.

Climate swings

Many plants and animals are very sensitive to the environment. Even relatively small temperature swings can wipe out selected species. Roger Lewin's article "Biologists Disagree Over Bold Signature of Nature" highlights some of the problems scientists face in searching for answers to mass extinction. George Stevens became keenly aware of a marked difference in the range of fauna and flora available in Alaska and Costa Rica after alternating between the two locations for several years. He is also aware of the species-rich gradient notion that recognizes a steady decrease in the number of species as a function of latitude from the equator to the poles. In Costa Rica, a single species is more likely to be limited to a very narrow band of elevation, where just the right environment exists for its survival. In Alaska, where survival requires tremendous adaptation to accommodate wide swings in environmental conditions between winter and summer, species can normally be found over a wide range of elevations.

During the winter of 1962-3, peach orchards in the Colorado River Valley between Fruita and Palisade, Colorado were hit by a cold snap that kept thermometers below zero for about three weeks. The deep freeze was not broken until warmer winds finally flushed out the pocket of cold air that was trapped in the valley. During the cold period, warmer temperatures were recorded on top of 11,000 foot Grand Mesa that overlooks the river valley. As a result of the cold snap, peach orchards on the Redlands District west of Grand Junction, as well as orchards in the Clifton and West Orchard Mesa Districts just east of Grand Junction, were nearly wiped out. Orchards in the Palisade and East Orchard Mesa District (at just slightly higher elevation) just a few miles further up the valley to the east were slightly better protected and suffered only scattered tree losses. Apple and pear trees lost the year's crop but the trees survived.

Swings in precipitation patterns can be just as destructive. During the dry dust bowl days of the 1930's, some plant life species were wiped out in the Great Plains. If the climate swing had been permanent, so would the regional extinction of many plants. Most of the lost species were, however, reintroduced into the region by neighboring regions where they had survived the drought. During dry weather periods, seeds may be preserved for several years. Indian corn has been reintroduced by planting and cultivating seeds found in ancient Indian ruins. Desert plants flourish after a wet spring, but their seeds may remain dormant for several dry years at a time.

According to one suggestion, ice age or near ice age conditions could result from massive outburst of volcanic activity with volcanic ash thrown into the upper atmosphere to cut off the sun's warming rays. Again some species would be able to adapt and survive, while others would find it impossible to hold on even for a few years.

Food supply. A secondary impact of climate swings is on the food supply. The food supply for many animals was restricted by the drought of the 1930's, causing a reduction in America's Heartland population, including that of humans. Carried to the extreme, the food supply loss causes heavy losses and even extinction of affected species. This includes the role of Nature's balance of predators. On the other hand, a species may be able to accommodate the climate swing, whereas its primary predator disappears. The surviving species may flourish for a short period, until Nature re-establishes a new balance.

The lower latitude flora and fauna species are more sensitive to climate changes than northern latitude flora and fauna. A minor axis shift could eliminate or critically reduce the available food supply for selected animal species. This domino effect amplifies the impact on some animal species.

Disease is just one more of the many variables that come into play in Nature's environmental balance. Plants and animals under the stress of environmental adversities become more susceptible to diseases. Added to the other variables, disease just may stack the odds so much against some species that they are unable to survive. Under normal conditions, Nature allows species to buildup resistance to diseases, unless the disease spreads through the population too rapidly. An axis surge, assumed to occur in a relatively short period, would limit the time available for flora and fauna to accommodate to new environmental conditions or build up resistance to new diseases.

Changes in sea level, even if small, can wipe out plant and animal life that is dependent on the tidal waters. It was the critical tidal zone that scientists used after the fact to determine the uplift of Mexico's west coast following the earthquake in 1985. Coastal lands were raised from only a few inches to three feet, resulting in the extinction of organisms that were dependent on the tides. All the scientists needed was to determine the elevation difference for the zones where tidal water species could no longer be supported. The same principle can be employed to determine subsidence.

Of course, the massive flooding suggested by sizable axis surges would drown many victims; first by the tsunami waves that raced inland, then by the permanent inundation of lands. Climate-sensitive species above the flooded lands would also be adversely affected by their new relationship to sea level.

Forest and range fires take their toll on plant and animal life by temporarily altering Nature's balance, but they seldom eliminate a species. In fact, the positive effects of wild fires are part of Nature's plan to clear land of excess vegetation and enrich the soils. The rebirth of vegetation often provides nourishment that helps species flourish. The fires, however, could trip the scales for species already in danger of extinction.

Combined stress. As with all things in Nature, it is seldom possible to pin the cause down to a single event. Each of the above factors contribute to the stress exerted on plant and animal life. When the sum of stresses exceeds a limit, something has to give. This can range from natural thinning and strengthening of some species, to extinction for other species. Since each of the above factors can be triggered by axis surges, the Dynamic Axis Theory could provide the answer.

Asteroid theory (dinosaurs' demise)

The dinosaurs' extinction has received the most publicity, even though they represented only a small fraction of the species lost by the same mass extinction event. Scientists place their extinction at the Cretaceous-Tertiary boundary about 65 million years ago. The debate continues on how rapidly they were eliminated. Louis Alvarez advanced one of the current headline theories. He proposed that the dinosaurs met their demise when the earth was struck by one or more of asteroids. The impact of an asteroid, he reasoned, would elevate a dust cloud into the upper atmosphere that would cut off the sun's warming rays and create a colder environment around the world. Increased traces of iridium found in the sedimentary deposits of that period has led support to the asteroid theory. Iridium, a product of meteorites, is a rare element on earth.

According to Bruce F. Bohor, Peter J. Modreski, and Eugene E. Foord more than 75 Cretaceous-Tertiary Boundary sites around the world have measured iridium anomalies that support Alvarez' theory. They added shocked quartz grains found in the same clays as further evidence of the asteroid theory. The association between iridium anomalies and mass extinction periods is not limited to the Cretaceous-Tertiary Boundary. The [K-T] boundary has received the greatest attention, partly because the dinosaurs were one of the prime victims, and partly because it is one of the most recent major mass extinction events. Consequently more clues are still available for study.

As with all science observations, it is worthwhile to simply reverse signs and take a look from the opposite perspective. Jack A. Wolfe, a paleobotanist with the Unites States Geological Survey in Denver, found evidence of greenhouse warming associated with the [K-T] boundary. To prove that any good debate has more than two sides, H. Jay Melosh of the University of Arizona, Tucson proposed that the asteroid strike would launch material above the atmosphere. When it fell back in a shower of glowing bits, they would have started wild fires around the globe.

A key unanswered question for the asteroid theory remains: if an asteroid strike could eliminate the dinosaurs, how could the dinosaurs evolve and survive over a period of millions of years without extinction? Surely an asteroid strike at the [K-T] boundary was not the only or most destructive strike during the life span of the dinosaur species.

Other scientists advanced evidence that the dinosaur extinction occurred over a much longer period. The debate continues over the possibility that large asteroids have struck the earth—for surely they have. However, were they the single cause, a major contributing cause, or was it just coincidental timing with the mass extinction event. Steven Stanley pointed out the lack of evidence for asteroids being responsible for any of a half dozen major mass extinctions. The only exception is the end of the Cretaceous Period some 65 million years ago. He noted that the Late Eocene Extinction may be tied to dramatic changes in the deep ocean currents. Of special interest is his version that the clockwise current circling Antarctica drives the cold deep ocean currents. A more detailed discussion of ocean currents is given in chapter 8.

The October 1990 issue of the Scientific American featured a debate on "What Causes the Mass Extinction." Walter Alvarez and Frank Asaro presented arguments for an extraterrestrial impact, and Vincent Courtillot argued for a volcanic eruption. Both articles included references from a long list of noted scientists that support the respective positions.

It is not within the scope of this book to attempt to enter the debate over the timing of mass extinctions, the selectivity of species wiped out in each extinction event, or the environmental impact of an asteroid striking the earth. The Dynamic Axis Theory does not need the asteroid theory to account for periods of mass extinctions. It also does not reject the asteroid theory as a contributing factor. From the perspective of this theory, an asteroid strike would add another contribution to the long list of weight shifts that affect the earth's dynamic balance. The resulting dust in the atmosphere provides another contribution to the short-term climate change process.

The Dynamic Axis Theory provides an alternate scenario for the mechanics behind periods of increased volcanic activity. Although it is difficult to prove that an increase in volcanism caused the extinction periods, it is equally impossible to prove that it had not major impact. When we add the influence of changing latitude and elevations to volcanic debris in the atmosphere, the ingredients for a mass extinction event are guaranteed.

Time scale. We need to keep in perspective the problems scientists face in trying to determine the time of the dinosaurs' demise. Assume the dinosaurs died out over a period of 65,000 years (hardly instantaneous), just 0.1 percent of the 65 million years since the event. To estimate the dates of a dinosaur's death within the 65,000 year period is equivalent to a forensic specialist's attempt to pin down to within 47 minutes the death of a crime victim killed 65 days earlier. If we include the additional loss of evidence associated with geologic time, the task of dating the death of an individual dinosaur becomes even more challenging.

The dinosaurs and other victims of mass extinction

For the Dynamic Axis Theory to be acceptable, it must also provide a safe haven during the axis shifts that occurred while the dinosaur herds evolved and populated parts of the earth for millions of years. Such havens would have existed along the neutral meridional and neutral equatorial great circles where little or no change in the geoid and climate occurred. Where climate changes are greater, species can still survive, if they either adapt or migrate to a more suitable climate.

The potential for any species to survive our hypothetical one-degree axis shift depends on the nature of the shift. The shift could involve a gradual axis drift, a series of smaller axis surges, or a single large axis surge. Any surge will have cataclysmic results with the degree of devastation being proportional to the axis surge size. It is reasonable to assume that the loci of the pole path over a series of small axis surges will move in the same general direction, as long as similar weight shifts continue to act. When the direction of pole movement approaches a straight line, the neutral regions will remain relatively untouched for millions of years, while dramatic changes occur along the shift meridians. Of course, the neutral equatorial great circle provides the best climate stability zone. It would be possible to identify many species threatened with extinction by the theoretical one-degree axis shift mentioned earlier, or any other theoretical axis shift. At the same time, species that would have a good chance of survival could be determined.

Although dinosaur fossils have been found in many areas of the world, we do not know that all areas were inhabited at the same time. If dinosaurs were living today, their habitat range (at least for some species) would be primarily limited to the swamp and bayou country of the United States Gulf Coast, Central America, Brazil, Central Africa, and Southeastern Asia where lush vegetation is available. Living in lowlands would make them more vulnerable to giant tsunami and geoid flooding triggered by an axis shift.

If the Louisiana Bayous were elevated, say 1,000 feet as called for by our hypothetical one-degree axis shift, much of the lush vegetation needed to support the dinosaurs would be lost for a time. Of course, a 1,000 foot drop in the elevation would drown all plant and animal life in the region. Consequently, surges will eliminate a species region by region. The luck of the draw would determine, if multiple axis shifts continue the extinction process around the world before the species has a chance to populate newly formed habitats.

If the shift meridian, for an axis surge based on today's land-ocean configuration, occurs along the 60 degree east and 120 degree west longitude, both neutral meridians (30 west and 150 east) fall primarily within the Atlantic and Pacific oceans. This would eliminate much of the neutral haven needed for survival. The neutral equatorial great circle would remain in place to repopulate other regions. The more limited a species' range, the more vulnerable it is to being eliminated by an axis surge. Even today we have a long list of animals on the endangered species list that can easily be wiped out by a minor environmental change.

It should also be pointed out that fossils of only a very small percentage of plants and animal life from past ages have been preserved for scientific study. Special conditions, some facilitated by a cataclysmic event, are necessary to preserve fossils. Consider the herds of buffalo that roamed the central plains. Multiply that number by thousands of buffalo generations over geologic periods. If even five percent of their bones had been preserved, we could not walk across a prairie without becoming entangled in their fossil skeletons. How about the billions of grasshoppers that invade the fields each year? They are soon gone without a trace. Even the evidence of last year's compost, that was used in the garden, is greatly diminished as Nature continues to recycle plant and animal life.

Consider also that most fossils are found on the surface of our planet. Underground fossils are excavated only when geologists have found a surface clue (including the surfaces exposed by excavation projects). Most of us would walk right over a fossil without giving it a second thought. Its existence will be recorded in scientific documents only when it is spotted by one of a few earth scientists trained in fossil recognition, or when it is brought to their attention. Even after discovery, we have a tremendous communication problem with one scientist relaying his or her knowledge and observation to other scientists. Obviously scientists cannot read all the literature. Consequently, all judgments and ideas of geologic phenomena are based on a glimpse at geologic history.

Much of our knowledge of past species derives not from fossilized bones, but from their tracks--similar to those that occasionally show up in freshly poured concrete pads. Here again, geologists can only study the exposed surfaces for evidence of the animals. Before a track can be preserved, it must be imprinted in a soft soil or mud, then solidified before it is destroyed. Unlike the stars at Mann's Chinese Restaurant, historic animals failed to leave their name and date along with their paw prints.

Periodic surges of the earth's axis provide a simple and logical explanation for the mass extinction events that have occurred during geologic history. Axis surges can account for either a single mass extinction event, when the species range is limited, or incremental extinction over a series of axis surges. Many of the suggested causes of mass extinction can be triggered by an axis surge, thus suggesting that each may be part of the Dynamic Axis Theory equation.

FROZEN MAMMOTHS

Among the many geologic mysteries are the frozen mammoths that have been found in northern Asia. The mammoths were unearthed with fresh vegetation in their mouths and with the digestive systems indicating their deaths during a growing season. Their meat was preserved for thousands of years in an icy tomb. For the mammoths to be preserved, they had to be quickly frozen within hours of their deaths before natural decomposition began. They had to remain frozen without interruption until their discovery. This would require a drastic and permanent climate change over a very short period.

Using a scenario that places the North Pole a few degrees south of its present position along about 50 degrees west latitude can account for a climate swing to trigger the last Ice Age. In that position North America and western Europe's higher elevations and latitudes would support the ice sheet, if air currents fed the areas with the required moisture. The same axis position would give the former Soviet Union a lower latitude and elevation before the shift that brought the pole to its present position. Land elevations, however, would have depended, not only on the geoid associated with its pole position, but on the crustal response to the higher magma seas. Obviously, the elevations of northern USSR could not have been substantially lower than today. Otherwise the area where the mammoths were found would have been under the seas. In any case, a warmer climate could have set the stage for grazing mammoths caught in the axis shift that suddenly converted a warm or moderate climate to an ice cold climate.

If the dating of the quick frozen mammoths can be correlated to the onset of the ice sheet meltdown, the same axis shift surge would explain both geologic events. Of the various geologic events contributing evidence for cataclysmic change, the frozen mammoths provide the strongest evidence that it was a very rapid event.

The preserved quick-frozen mammoths represent only a fraction of the expected loss of plant and animal life resulting from the indicated cataclysmic climate change. Many animals quick-frozen in that period must have thawed years ago under the present climate of northern Asia. When exposed, their carcasses decomposed as quickly as newly killed animals do today. Further south, there must have been more mass killings due to a climate change. Cold weather would still be one of the killing agents, but under conditions that did not provide the permanent icy tomb or lend themselves to fossilization. A sudden and dramatic climate change will result in mass killing and possibly mass extinction of particular plant or animal species. Other species will adapt and survive near the outer boundary of the affected region.

The Dynamic Axis Theory offers several factors that would contribute to the extinction of plant and animal life in different areas of the globe. The scenario of the rapidly frozen mammoths has a counter part in areas of the globe where warm climates replaced cold or cooler ones, or where dry and wet climates interchanged. The ability of prehistoric animals in North Africa to escape the arid conditions that triggered the formation of the Sahara would, in part, depend on how rapidly the change occurred. Migration to the south is a better possibility than adaptation to the barren Sahara sand. Satellite imagery has prompted ground research into a network of ancestral rivers in the Sahara. As for the demise of the mammoths, if the dates of the conversion of northern Africa into a desert can be correlated with the end of the Ice Age or another cataclysmic event, the Dynamic Axis Theory would receive a substantial boost.

Of all the clues of an axis surge, the frozen mammoths put the greatest time restriction on the event. The change in climate had to occur within a few days at the most.

Sedimentary Bedding

The distinctive layers of sedimentary beds offer the most complete record of a series of cataclysmic events. Consider the sedimentary beds now exposed for easy view by the erosion that carved the Grand Canyon and Colorado Plateau Region, and the logs from well drill cores. Under the right conditions, deeper and more spectacular Grand Canyons could be carved in many regions of all continents.

When the interstate highway (I-70) was cut through the hogback west of Denver, sedimentary beds covering many geologic ages were exposed. They had been tilted up at about a 75 degree angle when the Rockies formed. The cut immediately became a designated geologic point of interest. An interchange just west of the cut, exit 259, provides easy access to the parking areas on both sides of the highway. A short walk on a near horizontal walkway, several feet above the highway, makes it easy for visitors to walk through a segment of geologic time by viewing the changes in sedimentary bedding. One has a close-up view of the beds on this side of the highway. At the same time we can look across the highway to get a broader view of the same beds. The beds are colorful, changing between hues of gray, beige, tan, and maroon red, with hints of blue, yellow, and green plus traces of black that represent concentrations of hydrocarbon. The beds are only a few feet thick. They vary from course sandstone, that are part of the aquifer system supplying underground water for the eastern Colorado Plains, to layers of very fine grained shales, that serve as liquid barriers between sedimentary beds. Just over a mile to the South along Alameda Parkway, dinosaur tracks are exposed only a few feet above the highway. The discovery of dinosaur bones has given the area its name "Dinosaur Ridge."

About one-half mile to the west and further back in geologic time are the massive tilted red sandstone outcrops that serve as the backdrop for the spectacular Red Rocks Amphitheater. To the east is another mile of newer sedimentary beds that are nearly hidden in the bottom of a small valley and under the western edge of Green Mountain. Also to the east are tertiary sediments (including Green Mountain) deposited when the original Rocky Mountain uplift yielded soil, sand, and gravel to erosion forces.

Geologists use the distinctive bedding changes, that represent major changes in climate and geologic conditions, to establish their chronological calendar of geologic history. Each bed, or series of beds is assigned to a geologic age, epoch, period, and era. The changes in color, grain size, hardness, and mineral content are often dramatic between adjacent sedimentary beds. They would have required a radical change in either the climate, the source of sedimentary material, or the mechanics of deposition.

A review of today's accumulation of sedimentary beds will help us visualize some of the factors involved in the formation of geologic divisions. It also illustrates how the Dynamic Axis Theory can provide a logical scenario of events. All exposed land surfaces are subjected to both erosion and sedimentary deposition in varying degrees at the same time. Alpine topography and elevated barren plateaus yield the most sediments. Lake beds, flood plains, and oceans are the prime recipient of continental erosion.

Volcanic ash

Several inches of the volcanic ash from Mount Saint Helen's eruption coated the areas near the mountain. A gradation to lesser amounts of ash was deposited downwind from the eruption, as air currents circled the earth. Traces of the finest airborne debris from a large volcanic eruption will circle the globe several times before falling back to the surface. For the deposited layers of volcanic ash to be part of future geologic records, they must be protected by another layer of sediment or flood lavas to prevent them from becoming a victim of the erosion process. The dusting of volcanic ash is soon flushed off the hard rock outcrops of mountain ranges by heavy rains.

The John Day Fossil Beds National Monument in Oregon provides a prime example of how volcanic ash can preserve fossils of plants and animals. The first pages of the records contained in the fossil beds were cast about 40 million years ago. The abundance of fossils is attributed to the cataclysmic events of erupting volcanoes. The volcanic ash and mud flow buried the plant and animal inhabitants before they could escape to safer grounds. The cavity occupied by the trapped plants and animals were slowly replaced with minerals that preserved their size and shape.

One recorded history event that parallels the preservation of the John Day fossils by volcanic ash is the eruption of Vesuvius on 24 August 62 A.D. Herculaneum, a community on the slopes of Vesuvius, was buried under about 45 feet of pumice and mud. Pompeii was covered with 20 feet of ash. Again on 16 December 1631 Vesuvius came to life about 7 A.M. Six villages fell victim to lava, and nine to mud. Even Naples was covered by 18 inches of ash. Approximately 4,000 lost their lives. Archaeological exploration of the region started about 1,738. Coins, rings, and other artifacts were fair game for looters during the earlier years. In 1860, a more organized and serious archeological effort was undertaken. One of the big surprises was that the volcanic ash had provided an airless blanket that preserved some of Pompeii. Even though the bodies decomposed, the hard packed volcanic ash preserved their form as cavities—perfect molds that were filled with liquid plaster to reconstruct life-like forms. The cataclysmic event happened so quickly that many residents were trapped in natural poses.

Dust storms

Tons of sands and silt were eroded from the central plains states during the dust bowl of the 1930's. Most of the dust was redeposited on or near the same lands when the winds subsided—leaving fences, draws, and buildings buried by the drifting soils. A similar process has taken place in the San Luis Valley of Colorado, where the Great Sand Dunes accumulated in the northeast corner of the valley over a period of thousands of years. In New Mexico, the White Sands are different in color due to the difference of the sands' source, but with a similar mechanics of deposition. We usually think of the Sahara when we think of shifting sands, although smaller sand dunes are very common around the globe, including along ocean shore lines.

What will happen to the sand dunes now being formed during future geologic eras will depend on several factors. The sand is very vulnerable to stream erosion. If the White Sands area, for example, becomes elevated and tilted by future crustal gymnastics, the sands could simply be washed away. On the other hand, if the area was lowered to shallow seas or a lake bottom, new layers would be deposited over the sands. Today's sand dunes would become locked in as another page in the geologic history. If the sands contain or receive a cementing agent in the ground waters, the sand deposits can be bonded into a hard sandstone. A classic sample of past sand dunes can be found in the majestic formations of Zion National Park in southwestern Utah.

Land based sediments

Freezing and thawing cycles fracture and displace large rock fragments from mountain cliffs. Chemical decomposition of the bonding agents also helps loosen the rocks that fall from steep cliffs to form talus slopes at their base. Vegetation roots pry open small fractures in the rocks, making them more susceptible to erosion. Cloudbursts dislodge the loose silt, sand, and rocks to push or carry them along with the torrents of rushing waters. Alluvial fans are deposited at a sharp break in the slopes as the intermittent rushing streams lose their ability to carry the debris. Cloudburst torrents continue to inch along the large boulders, while bulldozing sand and gravel further down stream. Since a cloudburst is a short-lived event, the waters soon subside and most of the debris is deposited a relatively short distance from where it was dislodged. In arid country, such as the Great Basin, the next surge of the debris may be months or years later.

Some of the finer silts remain in suspension to be carried many miles before being deposited in a lake, reservoir, bayou, flood plain, or river delta. Much of the organic materials, from plant and animal life decay, also floats away with the stream. It supplies hydrocarbons that may eventually contribute to the oil and gas deposits of future generations. The gradation of sediments by size continues in the delta region as the bulk of the stream's sand and silt settle close to shore. However, some of the finer silts remain suspended to be carried by the ocean currents far from the delta into deeper waters.

Changes in Colorado River sedimentation

Before a series of dams was built in the Colorado River Basin, sedimentation concentrated at the delta in the Gulf of California. As each new dam was built, its reservoir became a settling basin for much of the river's sediments. We can use a one year cycle of sedimentation patterns in Lake Powell, behind the Glen Canyon Dam, as a miniature illustration of the cataclysmic changes indicated by sedimentary bedding.

Old timers in the irrigation farming country of western Colorado soon learned to recognize subtle differences in the color and texture of the Colorado River waters during the summer thunderstorm season. Each color and texture variation of the muddy waters provided a signature that helped identify the locality upstream where the storm clouds dropped their load. A particularly dark gray cast of very thick soupy waters told them that a cloudburst hit in the oil shale lands north of DeBeque and Parachute, Colorado. Cloudbursts in the Rifle to New Castle area turn the color of the sediment-laden waters of the river and canals to a lighter gray tone with a slight yellow cast. A bright maroon, thick, soupy water indicates that the storm hit the red beds of the Glenwood Springs and Redstone country. This was the river condition that prompted the comment by a pioneer of western Colorado that "the river was thick enough to walk across, but too thin to plow."

When the waters are less heavy with silt and the colors are less bright, a storm location farther upstream is indicated. The thinner texture comes from the fact that waters from the upper Colorado Basin drop some of its silt along the way. River waters are also diluted by clearer water tributaries as they progress down stream. Whenever erosion from different areas is combined, the color becomes a brownish gray. The gray-brown color is also typical for the spring high water runoff that represents a mixture of soils contributed by all tributaries.

When the muddy Colorado River waters enter Lake Powell, they soon drop their sediments, creating layers of soils that reflect their different sources. Interlaced with the upper basin sediments are the coarser sands and gravels deposited by cloudbursts in the Moab and Green River area just upstream from Lake Powell.

The deposits now being laid down in Lake Powell are the same as those accompanying a sea level rise to the surface of Lake Powell. A core sample of the sediments laid down in Lake Powell will reveal, not only the individual storms by their color and texture differences (mini cataclysmic events), but finer silts deposited during periods of clear running streams. The distinctive thin layer of winter deposition provides an annual demarcation in the varve, or sedimentary lamina. Geologists use it to establish short term geologic time scales in the same way that tree rings denote annual growth of a tree.

Before Glen Canyon Dam was built, the sediments carried by the Colorado River were deposited in Lake Mead behind Hover Dam. A core sample taken today from the upper end of Lake Mead sediments will show the cutoff of upstream sediments when the gates of Glen Canyon were closed. Lake Mead sediments, laid down after the Glen Canyon gates were closed, reflect the erosion that has taken place in the tributary basins between Lake Powell and Lake Mead. Should the gates of the Glen Canyon Dam be opened to drain Lake Powell, Lake Mead again would become the recipient not only for the annual upstream sediments, but also the softer sediments collected by Lake Powell that would be scoured out by cloudbursts and high waters.

Although the above scenario repeats as an annual cycle, the summer deposits illustrate how a charge in the source of sediments can create distinctive sedimentary bedding. On a much larger scale, the demarcation between the distinctive sedimentary beds around the world represents major cataclysmic events. Abrupt changes in fossil content between the beds document the magnitude of the cataclysmic event. We can see that climate swings are introduced simultaneously with the change in soil deposition.

The Grand Canyon—geologists open air laboratory

There is always more to learn from Nature's incision into the Colorado Plateau known as the Grand Canyon of the Colorado. Ron Redfern's book "The Making of a Continent", provides an excellent introduction to the Grand Canyon geology. Below all the sedimentary beds that make up the majority of the scenic canyon is 1.7 billion year old basement rock. The basement rock is now exposed where the Colorado River has cut down into Granite Gorge. The upper surface of the basement rock forms a platform on which sediments were deposited. The basement rock itself is Vishnu Schist, a Precambrian sedimentary deposit, interlaced with granite intrusions. The Vishnu Schist evidently formed during an earlier mountain building episode. The granite intrusions indicate that an insulating cover allowed the magma to cool slowly to form the coarse crystalline grains. Metamorphism in the Vishnu Schist also points out that sediments from an even earlier age were subjected to tremendous heat and pressure. The protective cap was needed to provide both insulation and pressure for this basement rock formation.

Whatever provided the insulation was completely eroded away before the first grains of the present Grand Canyon sediments were deposited. A geologic time gap of 1.13 billion years allowed plenty of time to strip the platform clear—much as the present Canadian Shield.

A series of sedimentary beds known as the Chuar and Unkar Groups were evidently the first deposited on the basement platform. After they were laid down, the region was raised and tilted. The elevated lands were peneplained by erosion to leave a wedge shaped block of the Chuar and Unkar Group sediments. This interruption in sedimentary deposition is known as the Great Unconformity.

The next sequence of sedimentary deposits laid down were the Tapeats Sandstone, Bright Angel Shale, and Muav Limestone. These episodes were followed by another unconformity erosion period when part of the sediments eroded away. The deposition of the remaining sedimentary beds that make up the Colorado Plateau was interrupted by no less than 12 more unconformity periods. Should the region again be lowered to below sea level, we would have to take a page out of the soap commercials to call it the super king size unconformity.

Many of the sedimentary beds were deposited in shallow seas. The crustal elevator had to therefore make several trips lifting and lowering the crust, while maintaining a near horizontal platform—except for the regional tilting during the Great Unconformity. Was it perhaps the crust that was being raised and lowered? Remember our discussion in chapter 2 about changes in the geoid. Would it not it be easier to explain the indicated elevation changes by axis shifts changing the geoid than a crustal elevator? We cannot get much help from the plate tectonics theory. It does not address the apparent bobbing action of the crust, except for plate collisions creating mountain uplifts with extensive crustal warping.

Within the overall pattern discussed above, we find that each distinctive sedimentary bed indicates a change in the sediment source or the mechanics of deposition. Many sediment beds also reflect a dramatic climate change. Something cataclysmic in scope must have happened to produce the dramatic changes between each adjacent sedimentary bed.

Even with the geologic evidence exposed for easy view on the Colorado Plateau, we do not know what, if any, sedimentary beds were on top of what we now see. We do not know the extent of each sedimentary bed that does not have outcrops on opposite sides of the Colorado River Basin. Nature systematically destroys more geologic records each year than has been documented in textbooks. Some estimates peg the sediment thickness for the Colorado Plateau before our present erosion cycle started at over 15,000 feet. Add to that the deposits eroded during periods of unconformity uplift, and we have a sizable sand and gravel pit.

More sediments

Have you ever seen the maps drawn to show ancestral invasions of the oceans at the time thick sedimentary beds were deposited and wondered about the sediment source?

They show a considerable reduction of land areas (no oceans converted to land areas). No mountains and high plateaus supply the volume of sediments laid down during the same geologic ages. How could such massive sedimentary deposits come from smaller continents?

The Grand Canyon is a very spectacular feature on a topographic map of the United States. The Colorado Plateau sediment thickness is not a prominent feature on a map ("Total Thickness of Sedimentary Rocks") prepared by Sherwood E. Frezon, Thomas M. Finn, and Jean H. Lister of the United States Geological Survey. The authors compiled this striking map from a wide variety of existing data. It uses contours to show the depth of basins filled with sediment.

Unfortunately, not all the available data were based on the same reference plane. Data along the Atlantic and Gulf Coast states, including the continental shelf, are based on the thickness as measured to the base of Mesozoic rocks. This leaves the probability that several thousand feet of additional sediments are present in some areas. Even so, many of the sediment depths dwarf the highest mountains.

One basin, over 45,000 feet thick extends from the coast at Mobil, Alabama, west through New Orleans, Louisiana then back onto the continental shelf to below Brownsville, Texas (the limits of the map). The deepest sediments are over 55,000 feet (over ten miles) thick at the Mississippi Delta. Another basin runs up the continental shelf from off the east coast of Florida to beyond Nova Scotia (limits of map). The deepest points along the east coast basin include:

· More than 42,000 feet at a point north of Grand Bahamas Island and east of Jacksonville, Florida.
· More than 30,000 feet at a point east of Wilmington, North Carolina and south of Cape Hatteras.
· More than 45,000 feet at a point east of Atlantic City, New Jersey.
· More than 50,000 feet at a point east of New York City, and south of Bangor, Maine.

Along the west coast, the depths are measured to the top of the Franciscan Series. The most prominent basin, over 20,000 feet thick, extends from about Taft, California north to Orland, California along the western edge of the San Joaquin and Sacramento Valleys. The thickest sections are 30,000 feet near the north end of the Sacramento Valley and 35,000 feet near the southern end of the San Joaquin Valley. Off the coast at Astoria, Oregon is a sedimentary filled basin more than 25,000 feet thick. A filled basin over 30,000 feet thick lies off the coast at Santa Barbara, California.

Information was not available for the sediments under the Columbia Plateau in Washington, Oregon, and Idaho, or for the Great Basin in Nevada, western Utah, and eastern California.

For the balance of the United States, the depths are given to the top of the crystalline basement rock. Some of the major features include:

·   Over 40,000 feet thick north-south basin from Salt Lake City, Utah into southern Idaho.
·   Over 25,000 feet thick Pecos Basin in west Texas.
·   Over 40,000 thick Anadarko Basin in west-central Oklahoma.

- Over 30,000 feet thick basin extending from eastern Oklahoma to Alabama. The deepest point is over 45,000 feet thick in western Arkansas.
- Over 15,000 feet thick circular basin centered in western Kentucky.
- Over 16,000 feet thick circular basin centered near Bay City, Michigan.
- Over 16,000 feet thick circular basin centered in western North Dakota.
- Over 30,000 feet thick basin running from southern west Virginia to east central Pennsylvania.
- Over 13,000 feet thick north-south basin under Denver, Colorado.
- Over 13,000 feet thick north-south basin under Cheyenne, Wyoming.
- Over 15,000 feet thick basin in north-central Wyoming
- Over 25,000 feet thick basin near Powell, Wyoming.

We have already touched on the geologic dilemma of how the crustal platform moves up and down between periods of deposition and erosion. The over 55,000 feet thick deposit at the Mississippi Delta multiplies the elevator range required to explain the Colorado Plateau sediments by a factor of about four. In the case of the Mississippi Delta, it could be argued that the sediments simply filled a depression in the class of the Mariana Trench with the added weight, responding to the rules of isostatic balance, pushing the floor even deeper into the magma seas. Such a scenario, however, is inconsistent with the formation of elevated horizontal sedimentary beds, such as the Colorado Plateau and Great Plains.

Where did the sediments come from?

A second and equally perplexing problem for geologists is to determine the source of such volumes of sediments. If we think of the sediment filled basins as being inverted mountains, their mirror image represents the mountain ranges that would have to be leveled by erosion to fill the basins.

Canadian shield

A possible contender is the Canadian Shield. Glacial erosion stripped off any sedimentary bedding that may have existed—leaving only the scratches and groves in the basement rock. Both granites and pillow lava igneous rocks are found on the Canadian Shield. The granites show that a thick protective cover provided insulation that allowed the magma to cool slowly when they were formed. The pillow lavas were cooled rapidly by ocean waters as they were ejected from the earth's interior. The exact sequence and scope of these events require detailed study by location and geologic dating of both the original basement rock and the dike intrusions. Some of the oldest rocks on our planet were found in the Canadian Shield Region of the North American Craton.

With the evidence of what ever provided the insulating cover destroyed, we can only speculate about its size or thickness. As mentioned earlier, the basement rock that supports the sediments now exposed in the Grand Canyon of the Colorado contains igneous and metamorphic rocks similar to the Canadian Shield. This similarity raises the obvious question, was the Canadian Shield at one time the foundation for massive sedimentary beds that provided some of the source material for the thick sedimentary beds found in the

United States? Of course, if thick sedimentary beds covered the Canadian Shield, they had to be formed by erosion of even earlier continental sources.

Ancestral Rockies

Geologic writings are dotted with references to the Ancestral Rockies—mountains that were leveled before the uplift that formed the present Rockies. No reference is available about the size of the Ancestral Rockies, but the sediments missing from the present Rockies are impressive.

The modern Colorado Rockies have apparently lost a two-mile thick cover off the top, in addition to the carved out valleys to erosion. This is based on a two-mile thick sedimentary bed pointing skyward along the eastern foot of the Rockies. The tilted sedimentary beds extend as near horizontal beds to the east, past the Kansas border. The same beds are found tilted up on the west side of the Continental Divide. It is a reasonable assumption that the outcrops were part of a continuous near horizontal sedimentary bedding that existed before the magma intrusion pushed up the modern Rocky Mountains.

The Colorado Plateau has lost all the sediments necessary to fill all the Colorado River Basin, including the Grand Canyon. Who can say how many additional sedimentary beds covered the region and were washed away? The Gulf of California is the recent recipient of eroded Colorado Basin sediments. The continental shelf sediments within the Gulf of California, however, fall short of accounting for all the sediments carved out of the Colorado River Basin and western slope of the Rocky Mountains.

The geologist's calendar of geologic history, as recorded in the sedimentary beds, provides the most complete chronology of the earth's cataclysmic events. The axis surges, fundamental to the Dynamic Axis Theory, address both the causes and effects indicated by the distinctive layering of sedimentary beds. The explanation, based on fundamental laws of physics and dynamics, is consistent with the scenarios developed for major climate swings, mass extinctions, crustal gymnastics, and the forces behind plate tectonics.

A challenging class project for our professor friend would be to determine how many erosion and deposition cycles have the average grain in the sedimentary rocks of the world passed. Each cycle mixes the sediments as the rivers collect debris from many hills and valleys, then does some sorting by size and weight before selecting its destination. Only a small infusion of new volcanic ash, meteoric dust, and chips off the igneous basement rocks are added to each new mix.

Submarine Canyons

More indication of an elevated continent comes from the submarine canyons that extend out from many of today's river basins. These include the Hudson Canyon cut down the slope of the continental shelf, the Laurentian Cone at the extension of the St. Lawrence River, and the Midocean Canyon between Labrador and Greenland. They all fit into the idea of a higher North America during the last ice age, or at some earlier period in geologic history. Additional submarine canyons indicate drainage from the Hudson Bay, Queen Elizabeth Islands area of northern Canada, and from all sides of Greenland and Iceland. Europe's major submarine canyons appear to drain the Baltic Sea, west-central Norway, and Lapland. From the map of the ocean floor published by the National Geographic Society, the near shore portion of the submarine rivers on the continental shelf appear to have been mostly filled by sediments after the geoid flooded the shelf. There is no surprise here.

Submarine canyons also exist along the west coast of North America, but with a much narrower continental shelf. They reflect the fact that smaller drainage basins feeding the western rivers were not fed by an outwash equivalent to the continental glacier that drained to the east and south. The western continental shelf is also associated with a much younger geologic feature, the Canadian Rockies, Cascades, and Sierras.

Evidence of a submarine river canyon can also be found in the North Sea, indicating river drainage of the Baltic Sea Region. Most continental land masses reveal evidence of drainage features extending well below any sea level accountable by ocean waters locked in glacial ice sheets.

Submarine canyons that extend out from most major river systems of the world, to depths as low as 10,000 feet, present interesting phenomena in geologic science. They exist, but since they do not fit into a favored working hypothesis, they are treated lightly in most geologic texts, and not at all in others. A frequently cited hypothesis assumes that changes leading to lower sea levels are primarily limited to the fluctuations in ice caps. Consequently, the range of sea level change can only account for about ±400-500 feet (even if we ignore isostatic balance adjustments). This is not nearly enough to account for the ocean depths of submarine valleys and canyons. Evidence of greater changes in sea level is usually attributed to crustal subsidence.

Earth scientists seem to readily accept that sedimentary rocks deposited below sea level are now near the top of the Himalayas and other mountain ranges of the world. They find it less likely that the submarine canyons now 10,000 feet below sea level were carved by conventional stream and river erosion while above sea level. Since the hypothesis limits the range of sea level change, it becomes necessary to search for a marine explanation.

Turbidity currents came to the rescue. The notion of density currents has scientific support. Cold, silt-laden melt water of alpine glaciers, flowing down the Rhone River into Lake Geneva, has been observed (through the clear waters) as it flows along the lake bottom to its deepest part. The turbidity current hypothesis uses this idea to explain how mud slides, on the continental slope, create currents that erode the submarine canyons and gullies. Breaks in the Trans-Atlantic telephone and telegraph line caused by mud slides are advanced as additional proof of their existence. Turbidity currents resolved another oddity. How else could one explain the transport of large sand and gravel particles beyond the lowest tides? How else could we rationalize coarse sands in the abyssal plains of the deepest ocean?

True, turbidity currents do exist. It is also true, that there are marine mud slides as the saturated banks of the submarine canyons give way to the gravitational forces. Tons of sediments cascade down the canyons toward the ocean deep. Why, however, would the canyons be aligned with existing river basins? The fresh water flowing into the ocean is less dense than the briny ocean. Thus, it does not provide a downward driving force. Silt-laden river waters will flow along the bottom, creating downward currents, until the sediments fall out of the solution. Observations of the Colorado River waters entering Lake Mead behind Hover Dam verify the current pattern. The currents, however, deposited their load in the old canyon rather than serving as an erosion force to carve a new canyon.

If turbidity currents are the eroding force of submarine canyons, the erosion pattern should have different characteristics than land based rivers. Submarine rivers must scour the canyon walls from the rim down, much like wind erosion, while surface rivers cut deepest into the river bed to form the classic "V" shaped valleys. The Hudson Canyon off the east coast of the United States is cut into the continental shelf sediments. Shelf sediments along the Hudson Canyon are easier to explain by turbidity currents than the Monterey Canyon off the west coast of California. The Monterey Canyon rivals the Grand Canyon of the Colorado and is cut into the harder granite bedrock.

The Dynamic Axis Theory offers a logical answer by providing variations in sea level elevations relative to the lithosphere. The theory permits a scenario for elevating any submarine canyon above sea level, so that the natural forces of stream and river erosion can provide the sculpturing tool. With the lands elevated, the existence of graded sand beds is no longer a big mystery. The Dynamic Axis Theory, of course, does not eliminate the contribution of the turbidity currents.

If we accept the idea that the submarine canyons could have been carved when the lands were above sea level, we have another indication of catastrophic changes in sea level. The process of inundating submarine canyons from the lower reaches to their present deltas would require millions of years of gradual sea level change. Silt from the river deposited at its delta would obliterate any signs of the offshore submerged canyon as it moved inland. The only way for a submarine canyon scar not to be buried in silt, therefore, is that an axis surge changed sea level by many feet. Higher resolution topography of the submarine canyons may be necessary to identify intermediate delta positions and associated marine terraces.

Fjords

Although we are on the topic of submerged crustal etchings, the fjords of Norway, British Columbia, southeastern Alaska, Labrador, etc. present another example where the erosion process requires greater sea level change than the vogue hypothesis allows. Fjords are the extension of "U" shaped glacial carved valleys. The erosion forces of glaciers are quickly reduced as the flowing ice enters an ocean. Fjords off Norway, however, have been carved to depths of 4,000 feet. As for the submarine canyons discussed above, fjord signatures would have been altered considerably, if sea level changes had been gradual. Also, assume the fjords of Norway were a product of the last glacial period. Imagine the difficulty of explaining their formation by ocean waters locked up in glaciers (a few hundred feet at the most) or by isostasy—specially since isostatic balance suggests that Norway would subside under the ice cap weight.

Magnetic Reversals

The earth's magnetic field has reversed many times during geologic history. This has been determined by checking the north-south magnetic directions locked into rocks when the magma first solidified. It was the matching of magnetic reversals as a function of age on both sides of the midocean ridge that provided the strongest arguments for tectonic plate spreading.

The causes and mechanics of the earth's magnetic field are not completely understood. Some new theories are entering the debate. Similarities with laboratory designed dynamos have provided some clues as reported by Kenneth A. Hoffman in an article "Ancient

Magnetic Reversals: Clues to the Geodynamo." We know that energy is required to maintain the intensity of the magnetic field. Otherwise it would die out in less than 10,000 years. Some scientists have suggested that the present magnetic field intensity is decaying at a rate that would reduce it to zero in about 3,000 years. It is now generally felt that in the reversal process the intensity reduces to zero before the reversal and then the intensity is rebuilt in the opposite direction. Once established, the magnetic field remains relatively constant for millions of years. No reversals have been detected for the Cretaceous era 135-65 million years ago. It would also be possible that a double reversal occurred so rapidly that the reversal record is missing or overlooked in the rock layers.

The magnetic field generator appears to be related to the mantle and liquid core currents. A slow westward drift of the field of about one degree of longitude every five years could be explained by a fluid core velocity of about one millimeter per second (100 yards per day). Centrifugal forces of the earth's rotation drive the horizontal circulation pattern, including the Coriolis effect. Heat engines in the earth's core drive the vertical circulation. Magma density, heat loss through the crust, and friction along the lithosphere and core boundaries must all play a part in driving and steering the magma currents and generating a magnetic field.

If the circulation pattern remains relatively constant, so should the magnetic direction. Conversely, if the magma current patterns changed dramatically as suggested by axis surges, could a dynamo reversal not occur? It has also been suggested that an earthquake shock above 7.5 could be the trigger for a reversal. The axis surges we have discussed will provide a series of crustal shocks (some could be near simultaneous) far above the suggested 7.5 magnitude earthquake.

R.T. Merrill and P.L. McFadden's article "Paleomagnetism and the Nature of the Geodynamo" provides a good mix between enthusiasm for the potential of unraveling the geodynamic mystery and caution as to how little is really understood. Their eye-catching statement from the perspective of the Dynamic Axis Theory, is a brief reference to a study of the 15.5 million-year period when basalt flows occurred in Oregon. Scientists are left scratching their heads by evidence of magnetic direction changes of a few tenths of degrees occurring in less than a year. Here again, an axis surge would alter the Coriolis forces acting on the magma seas in the lower mantle and outer core. The mixing patterns would be altered much like tilting a hand held mixer in a batter of marble cake dough. If the pattern of magma currents establishes the magnetic field, an axis drift could account for magnetic pole drift while an axis surge could account for the magnetic reversals.

The time frame for the magnetic reversal could be much greater than that for the axis shift. A relatively sudden axis shift would change the centrifugal forces acting on the magma. The magma current patterns, however, would take considerable time to change sufficiently to affect the earth magnetic field. This could account for a gradually reduction in magnetic forces to zero before it rebuilds the magnetic field with the opposite (or even the same) polarity.

If magnetic reversals are the product of a sudden jolt, the answer would appear to lie in the same axis surge events we have discussed. If the orientation of the magnetic field (position of the poles) is determined by the magma circulation patterns of the inner earth, data collected on magnetic pole movements may be helpful in plotting the magma sea currents by inverting the computation.

8   ODDS AND ENDS

Several loosely related geologic principles and events do not fit neatly into earlier chapters, nor do they justify a chapter status on their own. They do, however, contribute to the discussion of the role of axis movements in the evolution of our planet.

Milankovitch Cycles

A Yugoslav astronomer, Milutin Milankovitch presented a theory in the early 20th century to explain the climate swings of ice ages. Milankovitch's effort represents an extension and refinement of earlier ideas suggested by Joseph A. Adhémar in 1842. Adhémar speculated that astronomical events are behind ice age climate swings. Milankovitch used the cyclic variations in the geometry of the earth's orbit around the sun and variations in the tilt of the earth's axis to account for swings in seasonal intensities.

The earth's orbit oscillates between a near circular to a slightly more elliptical path over a cyclic period of about 100,000 years. The eccentricity of the planet's orbit varies from 0.005-0.06 (a circular orbit has 0.0 eccentricity). Although the variation in the distance between the earth and sun is small, it still has a token effect on the insolation (amount of sun's heat striking our planet). For today's slightly elliptical orbit, the earth is closest to the sun (perihelion) in January, and furthest from the sun (aphelion) in July. This calls for the Northern Hemisphere winters to be slightly warmer and summers slightly colder than for the periods when the orbit is nearly circular. The opposite is true for the Southern Hemisphere where slightly cooler winters combine with warmer summers. About 12,900 years ago, when our planet was emerging from the Ice Age, the earth was further from the sun during the Northern Hemisphere winters and closer during the summer. Thus, the insolation variation created by today's earth's elliptical orbit pattern was reversed. Precession causes the pointing of the earth's axis (extension of the axis above the north pole) to trace a conic path through the northern sky over a cyclic period of about 25,800 years. Precession results in a drift of the perihelion.

At present, the earth axis is tilted (obliquity) 23.5 degrees from the plane of earth's orbit around the sun. It is this axis tilt that is responsible for our seasons. When the earth is making its daily spins on one side of the sun, the tilt of the earth's axis allows more sun shine to strike the Northern Hemisphere. Summer weather north of the equator matches winter chill south of the equator. Six months later, when the earth is making its daily spins on the sun's opposite side, the earth axis tilt allows more sun rays to strike the Southern Hemisphere to reverse the seasons. Over a period of about 40,000 years, the axis tilt obliquity varies from 21.5 to 24.5 degrees. The greater the axis tilt, the higher the sun rises in the lands of the midnight sun (Arctic and Antarctic) to intensify the summer season.

This is offset by longer winter nights in the opposite hemisphere.

Milankovitch reasoned that the insolation may be sufficiently influenced by changes in the angle at which sun rays strike the earth (e.g., seasonal changes), and the distance variations between the earth and sun, to produce cyclic, global climate variations that could, in turn, lead to ice ages. The total radiation that reaches the planet from a constant sun varies only slightly. Milankovitch, however, calculated that the orbital geometry accounts for as much as 20 per cent annual variation in the higher latitudes of each hemisphere where ice caps are born and nurtured. He also recognized that periods of cooler summers, that melt less of the winter snows, combine with mild winters, provide the best potential for glacial expansion. One reason is that relatively warmer winters could also provide for more evaporation from oceans and heavier snows on the continents. This is the same reason why spring snows are often much heavier than midwinter snows.

For today's continental distribution, the Northern Hemisphere can provide the platform necessary to support additional ice sheet formation. Heavier snows in the winter months together with cooler summers could build an ice sheet on the northern reaches of North America, Europe, or Asia. Prevailing atmospheric currents and their moisture content become only one of the determining factors as to where and how much snow accumulates to form an ice sheet. The last Ice Age chose Canada, northern United States, and northern Europe, but skipped Asia and much of Alaska. Antarctica, nearly centered over the poles, is less affected by Milankovitch Cycles, since it lacks an additional platform to expand the ice sheet.

Each contributing parameter of the Milankovitch Cycle changes very slowly, and at different cycles. If the elliptical shape of the earth's orbit is the driving force for ice ages, they should occur every 100,000 years. If the tilt of the axis is the driving force, ice ages should come and go about every 40,000 years. If the combination of the above cycles teams up to cause an ice age, the greatest phase agreement would have a 200,000 year cycle, but with intermediate fluctuations between warm and cool climates. Add to the above cycles any contribution from the precession that causes the perihelion and aphelion to drift over a 28,500 cycle, and many more world climate fluctuations are possible. The resulting intensity of each phase agreement depends on the contributions of the individual cycles.

Based on plots of climate variations derived from deep ocean sediment evaluations, scientists suggest that the major ice age cycle period is about 100,000 years. They conclude that the earth's elliptical orbit cycle is the main driving force. Their climate plots show a gradual cooling with periodic interruptions of slight warming periods that continues for about 90 percent of the 100,000 year cycle. A rapid warming period that melted the ice sheets (except Antarctica and Greenland) completed the last ice age cycle in about 10,000 years. The sawtooth shape of the climate graphs of cold and warm periods demonstrates that the earth's orbital phase is not the sole cause of global temperature swings. If it was, the warming period would be closer to a mirror image of the cooling period.

Besides the geometric variations in the sun and earth system, we have variations in the sun's intensity. The best known fluctuation is the eleven year cycle of sunspots. It is now considered as only one aspect of a complex 22 year magnetic fluctuation cycle. As for many other natural cycles, the period and amplitude of the sun's intensity is irregular, opening the possibility that they may be the algebraic sum of several contributing cycles. One particular cycle known as the Maunder Minimum (named after E. Walter Maunder) has been proposed as the cause of the "Little Ice Age" during the 16th-18th centuries. As pointed out by Peter V. Foukal (30), scientists are still unable to accurately define long range solar cycles. They have not determined their correlation with Milankovitch or ice age cycles to establish global temperature swings over geologic history.

The idea of Milankovitch Cycles leaves earth scientists puzzled. The earth's orbit is responsible for the 100,000 year cycle and appears to provide the smallest variation in hemisphere temperatures. The elliptical orbit variation, however, is the only geometric variation in the Milankovitch Cycles that affects the total amount of sun rays striking the globe. This is the only cycle that directly addresses global, as opposed to hemispheric cooling and warming

If the very slight change in the elliptical shape of the earth's orbit is responsible for swings in climate that bring on ice and coal ages, it poses another interesting question. What is the chance for a planet in another solar system to have all our ingredients for life, including being at just the correct distance from the sun and having an elliptic orbit that would not destroy life completely at some period in its cycle?

Milankovitch's theory met with initial resistance and was generally disregarded for several years. Now the theory is finding some support in modern geochemistry. Scientists are using oxygen isotopes as clues to past climates. Fresh water locked up in ice sheets contains more of the oxygen-16 isotope. Thus ocean waters have a higher concentration of the heavy oxygen-18 isotope. Consequently, the ratio of oxygen-18 to oxygen-16 is a measure of the amount of ocean waters locked up in ice sheets. The O^{18}/O^{16} ratio cannot be directly determined for past climates. Instead, the calcium carbonate ($CaCO_3$) absorbed in shells of marine organisms on the ocean floor is used to indicate changes in the ocean waters. These, in turn, are assumed to reflect surface temperatures. The next step is to check for correlation between indicated cold spells (low seas) and the geometrically determined cycles of the Milankovitch theory.

Paleoceanographers feel they now have proof that cycle variations in the earth's orientation and orbit have controlled an approximate 100,000 year cycle of ice ages over the past million years. Franklyn Van Houten, of Princeton University, studied the 200 million year old Newark basin and found cyclic sequences of a full lake followed by low water that are close to precession and eccentricity orbit. Paul Olsen of Lamont-Doherty Geological Observatory also found cycle correlation in the Newark basin with relatively good agreement with computed 21, 41, 95, 123, and 413 thousand year peaks of the Milankovitch Cycles. In discussions of the Milankovitch Cycle, observed and computed values seem to vary between about 95,000 and 147,000 years. Also, if the Newark Basin represents lake bed deposits, the suggested cycle represents wet and dry regional cycles, not necessarily cold and warm cycles.

Scientists now suggest a super wet-dry cycle of 400,000 years. This super cycle could be explained partly by the phase agreements between the various components of the Milankovitch Cycle. The Dynamic Axis Theory simply adds another dimension in the influence of land-ocean distribution to regional climate. It suggests that similar cycles may be found at other locations on the planet that are out of phase with the Newark Basin record. In particular, any region that remains along the neutral meridional or equatorial great circles will experience less climate swings than for areas closer to the shift meridian.

The Milankovitch cycle theory received added support by Timothy Herbert of Princeton University and Alfred Fischer of University of California, Los Angeles. They checked 100 million year old marine sediments in central Italy and found good agreement for 100,000 and 400,000 year cycles. Testimony for 40,000 year cycles in 200 million year old marine sediments near Lyme Regis (along the southern coast of England) was found by Michael House of the University of Hull. It compared favorably to the tilt of the earth axis.

Additional questions and possible contributors to climate swings are given by Broecker and Denton's article "What Drives Glacial Cycles?" Their article cites evidence of near perfect agreement with Milankovitch Cycles using sea floor core studies conducted by Cesare Emiliani, University of Chicago and others. Their plots of ice volume (ocean waters locked up in glaciers) deduced from the O^{18}/O^{16} ratios indicate a cyclic buildup and retreat over a period ranging from 60-100 million years during the last 500 million years. For their cyclic period of 500-600 million years ago, the ice sheet did not increase in the sawtooth pattern of later years. It fluctuated around a volume midway between that of today and the height of the Ice Age. Geometrically driven Milankovitch Cycles, by themselves, call for a more uniform climate swing.

The use of oxygen isotopes as instrumental in unravelling geologic history. Oceans, however, are not homogeneous. We would not expect a core segment taken near the Azores that represents sediments laid down 100 years ago, to convey the same story about global climate as a similar core segment taken at the same latitude from the delta of the Ganges River. Why should we expect core samples from widely separated positions that were laid down 100,000 years ago to be more truthful?

Variation in the ocean waters locked up in ice during the peak buildup always depends on the available platforms in the polar regions. Local oxygen-18/oxygen-16 ratios depend on the influx of rivers, rain falling directly on the oceans, and the mixing action of ocean currents. With so many unknowns, interpretations of all drill core records must be tempered.

Winograd and coworkers reported (111) on a 250,000-year climate record. They found cyclic patterns that matched data from climate curves developed from Vostock, Antarctica ice core records and some marine records. However, the timing of the cycle phases disagreed by 7,000 and 17,000 years—a cycle shift. Of course, the dating differences have generated a heated debate—this time between the oceanographers and the terrestrial geologists. Coral records have since been introduced as a mediator between the deep ocean and terrestrial observations, as reported Richard Kerr (64). It is not within the scope of this text to enter the debate, except to introduce the Dynamic Axis Theory into the equation.

Because of the dating discrepancies between the different indicators and with theoretical Milankovitch Cycles, the Winograd group raised the question that orbital forcing may not have driven the Pleistocene Ice Ages (beginning about 2,500,000 years ago. Their study points out the difficulties earth scientists face in correlating geologic event dates. It also illustrates the danger of assuming that worldwide climates were somehow more consistent in the past than they are today.

How far can we trust geologic dating?

One of the critical elements in trying to reconstruct geologic records is the establishment of event dates. Tree rings represent accurate records, but they can only be used for the most recent events. Annual growth rings of trees have been used to date geologic events as old as 8,000 years. The oldest specimens, however, are very limited in numbers and global distribution. Sediment lamina, likewise, can provide accurate time scales where annual variations can be counted. Sediments deposited in an old lake bottom can have a near perfect lamina record. The record, however, has limited value until its oldest and youngest layers are indexed to other dated records. Any error in the indexing process can easily place the data in the wrong geologic age.

Radioactive elements have provided scientists with a dating tool to reach far back in time. Radioactive clocks are based on the half-life decay of radioactive elements. Scientists have established half-life rates of decay between elements that vary from seconds to thousands of years. However, we can never be sure of the absolute accuracy for the longer extrapolated half-life calculations. Are the determinations of decay rates accurate? Has fluid passing through the samples altered their chemical mix? Even though the accuracy of each test cannot be verified, they will continue to be used as the best educated guess.

Carbon-14 clocks, with a half-life of 5,700 years, are used for extending scientific dating up to about 40,000 years into geologic history. Scientists have questioned the absolute accuracy of carbon-14 dating for some time. Research at Columbia University established that carbon-14 dates in the 8,000-40,000 year range may be underestimated by nearly ten percent based on comparison dating with the more dependable thorium-uranium clock. The cause of the carbon-14 discrepancy is still debated, but it points out how easily accepted theories can mislead us. Beyond the fundamental debate of interpreting results of a single sample, multiple dating of similar samples often produces small variations. For example, if carbon-14 is used to date glacial advances and retreats, the sample selection presents a problem. Are samples taken from a terminal moraine a product of the latest glacial advance, or are they a piece of wood laid down during an earlier period that wound up in the heap of glacial debris? Dating new growth after the glacial retreat introduces similar pitfalls: is the sample from the first growth or later growth? For these reasons, it is not surprising that two competent scientists, using similar but different samples, can come up with conflicting evidence.

In theory, any radioactive element decay can be used to determine dates. For the terrestrial records in Nevada noted above, thorium and uranium isotopes were used. Argon and other isotopes are also regularly used. Each test represents a separate clock that must be calibrated with other dating techniques. For each sample questions remain. Was all the parent radioactive isotope present on day one, or was it a product of gradual accumulation? Only quick frozen magma provides a definite starting date. Leaching or other chemical reactions can compromise the sample.

Sedimentary rocks contain a conglomeration of materials, thus dramatically reducing their usefulness as positive measuring samples for radioactive dating. Calibration of sedimentary records is often accomplished by matching deposition patterns or common events that occurred in different data sets. This is fine as long as other variables do not come into play that contaminate one or more of the data sets. For lamina dating, eruptions of large volcanoes leave a distinctive signature that can often be detected on a global scale. Ash from smaller volcanoes provides additional regional signatures. Such records verify the accuracy of records obtained using other methods, as well as serving to index records like lake sediments.

Alternatives to Milankovitch Theory

Several serious questions remain. The Milankovitch Cycles are evolutionary (slow) in their response. Consequently, they do not provide the indicated trigger for the rapid decay of the North American and European ice sheets. The correlations, however, have been close enough to elevate the theory into much more serious consideration. Nevertheless, several attempts have been made to identify other driving forces that could account for sudden reversals in the ice age pattern.

Platform subsidence. William R. Peltier and William T. Hyde, of the University of Toronto, introduced the possibility of bedrock responses. They reasoned that as the ice sheet grew, the load might eventually reach a critical point that caused the bedrock to sink. The lowered platform would then contribute to the rapid meltdown. If this is true, should we expect the crustal rebound after the ice sheet decay to begin showing signs of a new ice sheet? The rapid sinking of the bedrock, suggested by Peltier and Hyde, applies similar arguments as the explanation for the gradual stress accumulation that is released by earthquakes. The Dynamic Axis Theory proposes the same reasoning that eccentric weights accumulate to a critical point before their release by an axis surge.

Atmospheric greenhouse gases. Evidence of carbon dioxide levels at about two-thirds the present level was found by Hans Oeschger (University of Bern) and Claude Lorius (Laboratory of Glaciology and Geophysics of the Environment) near Grenoble. If the sample represented anytime but the early stages of the ice sheet buildup, there is a small conflict with the logic presently being employed in the greenhouse debate. When the initial layer of the ice sheet was established, vast areas of forest and other vegetation in North America and Europe would have been eliminated. According to the proponents of greenhouse warming, a reduction of carbon dioxide-absorbing plant life elevates carbon dioxide in the atmosphere with a corresponding warming.

A fifty percent reduction in methane levels was also noted in the ice cores during the Ice Age. Methane, another greenhouse gas, is a product of vegetation decay, such as from swamp lands. The loss of swamp lands to produce methane, however, would appear to be the effect of a spreading ice sheet rather than the cause. If axis surges trigger an elevated level of volcanic activity, the carbon dioxide and other gasses released will alter the atmosphere mix, and possibly set in motion some greenhouse effect.

Atmospheric dust. Another condition noted in the ice age cores was dust at about thirty times the amount found in more recent snow deposits. Again, one can argue that the dust is a product of the surrounding lands that lost the more lush vegetation cover to frozen lands and tundra conditions. The question remains, did dust cool the lands by reflecting the sun's warming rays, or was dust simply a consequence of the glacial period?

Ocean Currents

Deep cold salty currents

In their search for a driving force for the global climate, Broecker and Denton came up with an interesting concept. Preferably, this force was triggered in some way by the 100,000 Milankovitch Cycle. Their concept of "Deep Salty Current" is a vertical and horizontal circulating pattern in the ocean waters. Warm and highly saline waters, indirectly related to the Gulf Stream, are presently cooled in the North Atlantic. This releases tremendous heat and warms western Europe. The denser, cold, saline Arctic waters sink under the warmer ocean waters to become the headwaters of a river, about 20 times larger than all the terrestrial rivers on the planet combined.

Besides the temperature-mediated density differences, areas of greater salinity are also heavier and sink below lighter ocean waters. A good example is the Mediterranean Sea. Its waters have a higher salinity than the oceans, and flow westward along the floor of the Mediterranean Sea through the Strait of Gibraltar, where they dive down to the ocean floor. Less briny ocean waters flow back into the Mediterranean above the out flowing saline current to maintain its sea level.

The deep salty current, proposed by Broecker and Denton as a trigger for climate change, flows down the west side of the Mid-Atlantic Ridge, crossing over the midocean ridge some place in the South Atlantic. Continuing around southern Africa, it crosses another midocean ridge between Australia and Antarctica before skirting Australia and heading for the North Pacific. For the waters to remain cold and dense, any crossing points on the midocean ridges must be inactive. Otherwise the extruding magma would warm the waters and disrupt the cold water flow pattern.

The deep currents' salty, cold waters evidently dissipated throughout the oceans leaving no comparable return river of warm surface flow. In some way, the cold waters are warmed to return toward the surface and eventually maintain the sea level datum of all oceans. Broecker and Denton proposed that changes in the deep sea current can account for peaks in the Northern Hemisphere Ice Age and contribute to global cooling. They developed a scenario, involving the drainage pattern of the North American ice sheet, to account for the termination of the Ice Age and a brief cooling event known as the Younger Dryas (a local cooling period in the North Atlantic after the meltdown of the ice caps were well under way).

Nicholas Shackleton of the University of Cambridge, England, presented evidence of cold waters sinking in the North Pacific at the height of the Ice Age 18,000 years ago. Cold salty waters diving into the ocean depth in the North Pacific would suggest a reversal from today's deep ocean current, or at least a dramatic change in its flow pattern.

The deep ocean currents addressed above cover only a small fraction of the climate's changes provided by ocean waters. Ocean currents play a vital role in establishing both regional and world climates. They directly affect atmospheric currents, thus extending their influence on climates. Oceans are credited with providing a leveling thermostat, since the range of temperatures (save over hot spots) varies from just under freezing in the Arctic Region to comfortable swimming waters near the equator. This is not close to the temperature range found in the atmosphere.

El Niño and La Niña

Scientists studying the equatorial waters of the Pacific Ocean are discovering how sensitive local climates are to relatively minor ocean current events. The subtle swings in the El Niño and La Niña (relatively warm and cold currents of the tropical Mid-Pacific) current patterns have been charged with causing drought in some areas, while at the same time causing heavy rains in others. If minor current changes of El Niño's magnitude can dramatically affect local climates around the world, imagine the climate impact of current changes resulting from the land-ocean distribution changes we discussed.

The Kelvin Wave mentioned in chapter V adds to the complexity of Pacific waters. Since the El Niño and La Niña currents cross the Mid-Pacific Ridge, it is worthwhile to see if their cycles correlate with heating provided by periods of peak magma discharge along the midocean ridge.

The Dynamic Axis Theory accepts most of Broecker and Denton's arguments. It also provides the mechanism to periodically alter deep salty currents by changing the land-ocean distribution and ocean depths. Other factors for ocean currents must also be considered.

After the earth has undergone a dramatic change in the land-ocean distribution, as the Dynamic Axis Theory proposes, the forces acting on the new ocean basins will dictate the prevailing current patterns. Centrifugal forces of our spinning orb employ the Coriolis effect to set up the horizontal circulation pattern. The size, shape, and depth of the ocean basins also dictate conditions that affect the ocean's circulation patterns. Ocean waters heated by the sun and by heat escaping from the earth's interior combine with Coriolis forces to create a complex pattern of vertical and horizontal currents and eddies. Since the heat is not constantly applied, the ocean currents are ever changing as the denser cold waters tend to sink as warm waters rise.

The Dynamic Axis Theory accepts the Milankovitch Cycles as a factor that affects long term climate changes. The correlations found in sediments are too strong to be ignored. As with most other theories, however, the Milankovitch Cycles theory is unable to stand on its own feet as the principle cause of ice ages. If Milankovitch's theory is accepted, it is necessary to develop another theory to address the indicated sea level changes of the magnitude needed to cause the marine terraces, submarine canyons and fjords as well as crustal gymnastics.

Supercontinents

Any discussion of the sediment sources of the geologic past must also address the plate tectonics ideas dealing with supercontinents. According to the continental drift theory, all continents were once combined into a single continent called Pangea. The thick sediments covering the United States and its continental shelves, therefore, could have come from the lands of other continents. On the other hand, it can be argued that the North American continent gave up about the same amount of sediments as it received.

According to the theory, Pangea split into two supercontinents of Gondwana and Eurasia, which in turn divided into the continents as we know them today. Continental division is still in progress. East Africa's Great Rift Valley appears to be in the process of splitting the Horn of Africa (Somali, eastern Ethiopia, Kenya, Tanzania, and Mozambique) off the rest of the continent. Some of the rift valley is already well below sea level, but remains dry since the Danakil Alps block the entrance of ocean waters from the Gulf of Aden and the Red Sea.

Presumably, at some later date in geologic history, today's divided continents will recombine into one or more supercontinents. This alternating episode of joining and splitting continents is often referred to as the Wilson Cycle. Building on the theory of Pangea—Nance, Worsley, and Moody pooled their respective specialties of tectonics, oceanography, and geochemistry to formulate a theory for cyclic supercontinent formations. From their data they have identified evidence of five cycles of supercontinents beginning about 2,600 million years ago as explained in their article "The Supercontinent Cycle." Each cycle covers a period of approximately 500 million years. The driving force for their proposed supercontinent cycle is the heat produced by the decay of radioactive elements at the earth's interior. They accept the conveyor belt concept as the driving mechanics for tectonic plates. Central to their theory is that about half as much heat is conducted out through the continental crust as through the ocean basins. (Remember our earlier discussion about the isothermal-pressure balance and the Moho?) This means that heat is trapped under a large continental crust. They suggest that eventually the trapped heat torches its way up into the land mass to create rift zones that split the continent apart and form new ocean basins. In this way, they suggest, a relatively uniform heat source can be released in sudden bursts to account for episodes of mountain building and glaciation.

The breakup of a supercontinent continues until the lands are too small to trap the heat. When the heat is dissipated by their surrounding oceans, the continents move back together to complete their cycle. Not all lands and oceans are affected identically. For several of the supercontinent cycles, apparently the Pacific Ocean has remained a near permanent ocean, while the Atlantic Ocean as periodically opened and closed. Analyzing the geologic evidence of mountain building and plate movement rates, the Nance group calculated the supercontinent cycle at about 440 million years—a little shorter than the earlier noted 500 million years.

They also addressed the periodic flooding of continental lands as a product of the supercontinent cycle. Low seas, they suggest, exist when deep oceans surround the combined continents. As the new ocean opens between continental sections their seas are shallower, consequently sea levels are higher. Today we find the Atlantic Ocean crust moving out from the midocean ridges. The cooled ocean floor becomes more dense causing it to sink to create a deeper ocean and lower the seas. The Nance group article claims good correlation between the proposed supercontinent cycle dates over the past 570 million years and changes in sea level. It does not, however, address the sudden changes in sea level required to account for marine terraces as discussed in chapter 7. It is also interesting that the theory considers changes in sea level completely unrelated to the amount of ocean water locked-up in continental glaciers. Part of this can be attributed to the time scale under consideration. The Nance group briefly addressed the cyclic aspects of ice ages and life forms with supercontinent cycles and suggested that the supercontinent cycles are dominant.

Supercontinents and the Dynamic Axis Theory

A vital part of the supercontinent theory is the assumption that all the continents have remained essentially the same. The only exception is the accretion of island terrains during plate subduction under the edge of the continent. The heart of each continent is the craton which is promoted as a permanent land mass. The continents are assumed to have retained their form with only relatively minor interchanges between continental land masses and ocean basins. Maps of Pangea, Gondwana, and Eurasia account for all present continents with only minor alignment changes in their shore lines to force fit their jig-saw pieces. The continents are correctly assumed to have occasionally invaded and molded the temporary continental shelves, while adding new sedimentary beds. The theory also holds that whilst the continental boats were sailing the magma seas, the oceans' crustal floors were continually recycled. New ocean crust formed along midocean ridges where the plates were spreading. An equivalent amount of ocean crust was consumed along the subduction zones. The finding that dated basement rock of the ocean crust is generally much younger than dated continental basement rock provides strong support for this idea.

This frequently cited simplistic explanation, however, fails completely to recognize the discussed changes in sediment-filled basins within the continents and along the continental shelves. The problem arises in part from the hypothesis that horizontal forces are the primary driving forces acting on the crust (magma currents driving tectonic plates). Vertical movements of the crust become a secondary reaction of plates in collision, as proposed by the tectonic plate theory.

The Dynamic Axis Theory reverses the process by assuming that centrifugal forces resulting from axis movements produce vertical forces which act on the crust. (It takes much less horizontal force to slide a large plywood sheet, if one corner or side is raised first.) Horizontal and shear forces become secondary reactions as the crust must stretch and compress to fit the ever changing magma sea ellipsoid. The Dynamic Axis Theory also provides for elevation (sea level) changes with or without uplift or subsidence of the crust. It simply attributes these to centrifugal forces acting on the planet's oceans.

The Dynamic Axis Theory accepts the notion that land masses join and divide and that heat from the earth's interior plays a vital role. The primary weakness of the supercontinent cycle theory is its dependence on conveyor belt mechanics to propel the plates. Its slow evolutionary rate of change fails to consider the cataclysmic events discussed in chapter 7.

Atmospheric Pressure and Earthquakes

Jerome Namias is a meteorologist at Scripps Institution of Oceanography, in La Jolla, California, and a former Chief of the National Weather Service. He has found some correlation between unusual high atmospheric pressure patterns over the eastern North Pacific Ocean, low pressure over the western United States, and the occurrence of moderate to large earthquakes. In particular, he postulated that the southern California earthquakes in July 1986 and October 1987 were triggered by changes in atmospheric loading on the crust. The correlation is weak and the concept is a very unusual suggestion, according to Namias. The earthquakes tend to occur between one month to six weeks after the onset of the unusual atmospheric pressure patterns. Namias' speculation involves high winds associated with the atmospheric conditions raising offshore sea levels, thus changing the loading weights on the crust. Wind driven surf is similar to tides produced by the moon, but not as great in magnitude as. Lunar tides were suggested to influence earthquakes long before this book proposed that they combine with magma tides to massage the crust.

Namias later expanded his study to include the period from 1947-1987. He divided the summers into groups with many or few earthquakes in excess of 4.5 magnitude. The correlation of summers with atmospheric conditions provided additional support for his earlier findings.

Differential loading on opposite sides of the junction of tectonic plates, where a natural weakness already exists, could be the straw that made the difference in overcoming frictional locks between plates. This is specially true for some of the dip-slip faults in the Los Angeles area. The Dynamic Axis Theory accepts Namias' idea as one of many contributing force differentials that are applied to the earth. It treats it more as an indirect than direct cause of earthquakes.

It is interesting to carry Jerome Namias' speculation one step further. Analyzing his suggestion with the Dynamic Axis Theory affords a slightly different perspective. We need to dust off Kelvin's suggestion that atmospheric high and low pressure cells contribute to the ever shifting weights on our planet and give it a slightly different twist.

As high and low atmospheric cells tend to move from west to east across the mid-latitudes of the Northern Hemisphere, they shift the weights on the planet like a convoy of vehicles crossing a bridge. Fig. 8.1 shows a hypothetical string of high and low pressure cells in the pattern similar to that suggested by Namias, with high pressure in the eastern Pacific and low pressure over the western United States. Our hypothetical pattern extends the convoy of pressure cells. First, we limit one of the continental low pressure cells to the area between the San Andreas and Wasatch Fault zones. Second, we add a high pressure cell

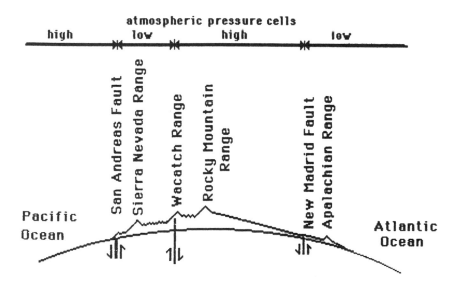

Fig. 8.1 Hypothetical atmospheric loading on oceans and continents.

between the Wasatch and New Madrid fault zones. Third, we add another low pressure cell over the eastern United States. The fifth zone, high pressure over the western Atlantic is not shown.

Atmospheric cells over ocean waters

The effect of variations in the atmospheric pressure patterns are quite different for oceans than land surfaces. Over the oceans, the high pressure cells will displace the ocean waters like a huge boat. The displaced ocean waters are driven to areas of low pressure cells (isostatic balancing). The high tide associated with the low pressure cell of a hurricane is a good example of the hydrosphere compensating for changes in atmospheric pressure. The weight of water displaced by a high pressure cell should approximately equal the weight of the high pressure cell that exceeds the mean atmospheric weight. Over the open oceans, therefore, the loading of the planet should remain in near balance as the sum of atmospheric and ocean waters tend to remain essentially constant. In the process, however, some ocean waters from wind generated tides will become trapped against land masses and produce a weight abnormality. The vectors shown in Fig. 8.2 reflect the atmospheric loading that pushes the crustal barge downward under high pressure cells, without consideration for displaced ocean waters, and the upward rebound under low pressure cells.

Atmospheric cells over continents

Over land, atmospheric cells act like cargo on the crustal barge floating on the magma sea. High pressure increases the load and pushes the crustal barge deeper into the magma sea. Low pressure unloads the crustal barge and allows it to float higher in the magma sea. In Nature, the full affect of changing atmospheric loading is probably never realized because of the viscosity lag of the magma seas.

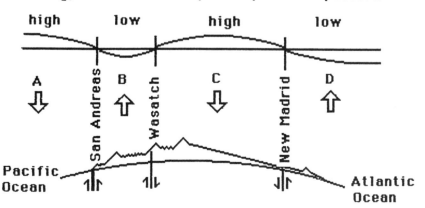

Fig. 8.2 Differential shear at fault zones due to atmospheric loading.

Differential shear at fault zones

The differential shear created at the fault zones addresses Namias' suggestion. Consider the weight changes associated with a high pressure atmospheric cell over the Pacific, as Namias suggested. For this oversimplified scenario, assume each block of crust is rigid and floats on a magma sea. Any shear developed within the crustal block is transferred to the block edge (fault zone). As shown in Fig. 8.2, the Pacific Plate [A] is pushed down into the magma sea as indicated by the downward vectors on the Pacific side of the San Andreas Fault. The low pressure atmospheric cell over the crustal block between the San Andreas and Wasatch Fault zones [B] unloads the crustal barge. These upward force vectors are shown on both sides of crustal block [B], and indicate the barge is being elevated. A vertical shear force is created along the San Andreas fault zone. Since the crustal barge between the Wasatch and New Madrid Fault zones [C] is pushed down by high pressure, whilst the crustal barge [D] east of the New Madrid Fault zone is pushed up, vertical shear is also created along the Wasatch and New Madrid fault zones.

At this point in our analysis, we have a vertical shear force acting on the existing faults. Any responses along the fault will depend on the shear force overcoming the friction lock. The atmospheric loading is not enough by itself to trigger a rupture, but could combine with other forces to accomplish this.

The magma tide component

If the crust is also under the influence of a positive quadrant magma tide when the atmospheric cells are in place, its higher magma pressure introduces an upward force vector. The shift meridian position (longitude) determines the relative shearing components at the fault zones. When the crest of the positive quadrant magma pressure (shift meridian)

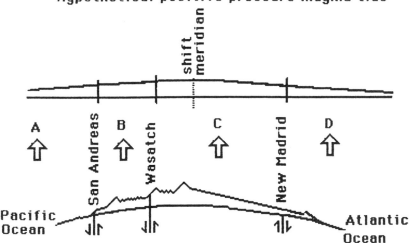

Fig. 8.3 Differential shear at fault zone due to magma tide.

is east of a fault zone, the resulting differential loading is greater for the east side of the fault than on the west as shown in Fig. 8.3. The San Andreas and Wasatch fault zones are assumed to be west of the shift meridian (magma tide crest; Fig. 8.3). For this illustration, the shift meridian is west of the New Madrid fault zone. It therefore provides a shearing differential that has an upward vector on the west side of New Madrid. The magma tide, under this theoretical condition, increases the shear forces along the San Andreas fault zone, while decreasing the shear force along both the Wasatch and New Madrid fault zones.

An additional and more critical factor associated with the positive quadrant is crustal tension from magma pressure. With the crust under tension, the existing faults tend to pull apart. This reduces the friction lock and reduces the force required to rupture the crust. The potential for rupture along a fault therefore becomes a function of the differential loading on each side of the fault. The crust rigidity transfers the crustal loading differentials to the fault zone (any flexing of the crust under load reduces the effective stress at the fault surface), and the existing friction locks.

If the crust is under the influence of a negative quadrant, the lower magma pressure simply reverses the direction of the force vectors. Again, the position of the shift meridian determine the relative force vectors introduced along the fault. The vertical force contributed by a negative quadrant of a magma tide reduces or possibly reverses the shear forces created by atmosphere loading acting on the San Andreas Fault. Reduced magma pressure increases the compression forces, thus increases the friction lock on near vertical faults and reduces the chances of an earthquake. For low angle dip-shear faulting, however, the added compressive force could increase the potential for rupture.

The magnitude of the forces resulting from atmospheric weight shifts by themselves appear to be too small to cause fracturing along the fault zone. Nevertheless, they could provide the nudge to trigger a rupture in concert with all other forces that act on the crust. Examples include magma tides from the Chandler wobble, moon and sun tides, axis drift, and wind tides. Compression and tension forces generated by tectonic plate movements, of course, are prime factors. In the specific case of the San Andreas Fault, the primary stress is the strike slip caused by the Pacific Plate's northward movement from to the North American Plate. The vertical shear forces illustrated above can initiate relative plate movement that breaks the friction lock and results in a large horizontal earthquake movement. Dip slip faults created by the collision force component between the Pacific and North American plates, should be more responsive to direct variations in atmospheric and magma loading (see Appendix D for more detailed analysis of forces acting on fault surfaces).

All crustal faults are affected by the massaging action of atmospheric loading, magma tide pressure variations, lunar- and solar-driven tides, or any other variable with cyclic patterns that alters the lithosphere loading patterns.

The four to six weeks time lag Namias noted between the buildup of high pressure over the eastern Pacific and earthquakes along the San Andreas Fault zone remains a mystery. Movements of high and low pressures normally will completely change the loading pattern over a period of a few days to a couple of weeks. Near stationary or slow moving atmospheric cells provide more time for isostatic balancing adjustments between atmosphere and ocean waters. Isostatic balancing between the crust and magma seas takes even more time to respond. The time lag between the buildup of pressure cells and crustal rupture, therefore, may simply represent the delay in hydraulic response time of the magma seas to the differential atmospheric loading.

Tug-of-war

The atmosphere plays the same role for earthquakes as the class weakling did in a high school tug-of-war contest. The setting was a ten foot wide and three foot deep irrigation ditch. The schedule called for the war to begin at high noon (during lunch hour). The contestants were the high school freshmen against the sophomore boys. At stake was the right of the sophomore class to impose a long list of initiation rules on the freshman class. The sophomores chose the north side of the ditch with some weeds and grass, but softer soil that they felt could provide better footing by digging in their heals. This left the freshmen with the south side with a hard pack dirt road on a slightly elevated ditch bank.

During the first three minutes of the contest, neither side gained an advantage. Slight movements back and forth occurred as one or more of the contestants' feet would temporarily los their friction lock. The freshmen class weakling showed up late to join the contest. Ever so slowly, the advantage began to shift toward the freshmen's side until the first sophomore slipped into the canal. The freshmen continued to consolidate their gains, as one by one each of the sophomores hit the water.

Headlines in the school newspaper declared: "Freshmen Weakling Dunked Sophomores." The freshmen, however, gave the primary credit to their 230 pound anchor. Atmospheric high and low pressures in the right place, at the right time may just tip the balance of forces in the same way as the class weakling made the difference. Of course, in cataclysmic earth science events the Milankovitch Cycles, an asteroid strike, volcanic eruptions, or greenhouse gasses have each been promoted at some time as the primary cause for climate swings.

Crustal Gymnastics and the Mediterranean Sea

There is evidence that the Mediterranean Sea includes experienced at least one cycle from a deep sea to a dry basin before returning to its present condition. Evaporates found between two layers of deep sea deposits on the abyssal plains about, 10,000 feet below sea level, provide the clue. It appears that the ocean waters blocked by the Isthmus of Gibraltar turned the Mediterranean Sea into a 10,000 foot deep dry valley.

Kenneth Hsü's scenario calls for the Ancestral Mediterranean Sea being formed out of the Tethys Ocean by continental movements after the breakup of Pangea. When the west end of the Mediterranean Sea was blocked off the Atlantic, its waters evaporated in what he called the Messinian salinity crisis. Fresh water rivers feeding into the basin, like rivers feeding the Great Salt Lake Basin, were more than offset by evaporation. The rivers, however, cut deep canyons that have since been filled with sediments. The Russians discovered a gorge under the Nile Valley near the Aswan High Dam at a depth of over 600 feet below sea level. The deepest portion of the Ancestral Nile Valley is filled with about 400 feet thick marine sediments. Above the marine sediments are about 500 feet of freshwater sediments, bringing the Nile floor at the dam to over 300 feet above sea level.

Similar evidence of deep stream erosion near the delta of the Rhone River, now buried by sediments, provides further support for the argument of missing waters in the Mediterranean. Additional support derives from the river silt, sand, and gravel found far from today's shores, and at depths normally coated with thin a thin layer of ooze.

The abrupt change in sediments, found by the sea floor drilling in the Mediterranean Sea, led Hsü and his colleagues to the conclusion that the Gibraltar dam broke. This allowed ocean waters to enter the Mediterranean Basin over a fall that would make the Niagara look like a brook by comparison. They also postulated that the Strait of Gibraltar may again block off the Atlantic waters and return the Mediterranean to a desert basin.

The Dynamic Axis Theory brings more options to the scenario. If the region was uplifted by higher magma seas, it would not be necessary to visualize the evaporation of a 10,000 foot deep basin that left only scattered pockets of brackish waters. If the region's elevation was higher on the eastern end, much of the ancestral Mediterranean Sea waters would have simply poured out the Strait of Gibraltar. This would dramatically reduce the water volume that needed to evaporate. Remember our discussion in chapter 4 about the tilting of the Great Lakes Region as a result of a hypothetical one-degree axis shift. A west-flowing falls at Gibraltar, with a submarine canyon extending into the Atlantic, would provide evidence that the flow through the strait was a two way canal.

Texas Wave

Thor A. Hansen, a geologist at Western Washington University, reported on data for an approximately 300 foot high wave that crashed into the Texas coast about 66 million years ago. The coast line was much further inland at the time, making the present Texas coast line part of the Gulf of Mexico. The evidence of the massive wave includes the displaced suitcase-sized blocks on the ancient ocean floor, the deposition of sand and shark teeth in the Brazos River area of south-central Texas, and the scouring of the ancient Gulf of Mexico floor.

Mr. Hansen theorized that a huge asteroid hitting the earth in the Gulf of Mexico caused the wave. Iridium found in the sedimentary layers in conjunction with the wave deposits, provides the clue that pointed to an asteroid impact. This resembles Louis Alverez's theory on the cause of mass extinction of dinosaurs. The difference is that the asteroid Hansen considered struck in the Gulf of Mexico, rather than on land where it would thrust dust and debris into the atmosphere and stratosphere.

Eric Kauffman of the University of Colorado suggested the asteroid strike is still tied to the extinction event that eliminated dinosaurs from earth. The contribution to extinction from an asteroid striking the ocean, he observed, "comes from a change in the earth's temperature, water chemistry, and climate." He noted that the Texas tsunami appears to be part of a series of short-term, closely spaced extinction events that included a meteor whose 25-mile-wide Manson Crater was discovered in Iowa.

Two competing teams provided additional data apparently relating to the same asteroid strike. The first group was that of geologist Bruce Bohor of the United States Geological Survey and material scientist Russell Seitz of Cambridge, Massachusetts. The second group was that of planetary scientists William Boynton and Alan Hildebrand of the University of Arizona. Evidence found in Haiti, Cuba, and Alabama is added to the Texas data. Bohor and Seitz suggested an impact site south of western Cuba, whereas Boynton and Hildebrand proposed a site off the coast of Columbia.

Frank Kyte and two colleagues of the University of California at Los Angeles suggested that an asteroid strike in the South Pacific may have contributed to the onset of the last glacial age some 2.3 million years ago. The asteroid, estimated to be about a quarter-mile across, was discovered by studying core samples from six sites southwest of South America, near Antarctica. Kyte did not propose that the asteroid strike directly triggered the Northern Hemisphere glaciation that occurred about the same time. However, he suggested that water thrust into the upper stratosphere reflected some of the sun rays back into space, thus cooling the planet.

The Dynamic Axis Theory does not disagree with the notion of asteroid strikes contributing to cataclysmic geologic events. The Texas Wave evidence cited by Hansen can also be explained by a very minor axis surge. It is the iridium anomaly that points the finger at an asteroid strike. This raises the question: does other evidence equivalent to the Texas Wave exist without an iridium anomaly?

Internal Plate Extension, Compression, and Rotation

According to some reports, alternating periods of extension and compression are found in the crustal formations along the eastern edge of the Great Basin. The Cordilleran fold and thrust belt studied extends from southeastern Idaho down across Utah, southern Nevada and into California. This is part of a fold and thrust belt that is discontinuously exposed from Alaska to Mexico. Scientists report a 65-83 mile shortening of the basin some 150-50 million years ago. There is also indication of about 153 miles of basin extension (mostly east-west) about 37 million years ago.

Relative crustal movements across a fault boundary have been much easier to establish than absolute movements. For help on the bigger picture, scientists are expecting an increase in the density of a VLBI geodetic control network. To solve the puzzle of the Great Basin's formation, it is necessary to track the translation and rotation movements of the North American Plate. It is also essential to determine the internal shortening, extension, and rotation that occurs within the Great Basin and for other relative regional movements that may exist. Signs of crustal gymnastics suggest a non-uniform stretching and compression within the plates. The internal stretching of the Great Basin is also a factor in resolving the discrepancies between the observed slip rate along the San Andreas Fault and the indicated plate movement rates.

The answer may be partly found in the internal plate deformations in the Coastal Ranges and the Basin and Range Province of the Great Basin. Jordan and Minster reported on the use of VLBI data to establish present day extension rate estimates for the Great Basin. Their results generally agree with estimates derived from geological observations. The California Institute of Technology, the Massachusetts Institute of Technology, the University of California at Los Angeles, the University of California at San Diego, and several Government agencies set up a new geodetic network in 1985, designed specifically to obtain the needed data. Each new set of observations will refine internal plate movements for the region—or add to the complexity and confusion.

The Dynamic Axis Theory accepts the role magma currents play, under the thinner crust of the Great Basin, in the crustal extension compression and rotation. It also provides an additional explanation for crustal shortening and extension as a function of great-circle half-lengths resulting from axis movements (see Table B-2). To account for the indicated magnitude of shortening and extension within the Great Basin, however, requires additional stretching and compression, possibly related to local forces generated by magma intrusions and subducting plates. When the mechanics of the Great Basin is better understood, we will surely apply the concepts to many other regions of distorted crust within other tectonic plates.

Ultradeep Drilling

The Dynamic Axis Theory may have the answer to new questions raised by deep drill holes, such as the Kola Peninsula Project (near Scandinavia) in Russia, and the Oberpfälz Forest Project in northwest Bavaria, West Germany. The Kola Peninsula drill hole (one of eleven deep drill holes in Russia) has reached a depth of 39,586 feet (12.066 kilometers) of its target 49,200 feet (15 kilometer) depth. Since it was drilled two decades ago, scientists have had time to digest some of the findings. They found nickel and copper ores at about one mile depth. They also found appearances of copper, nickel and lead deposits at 20,000-36,000 feet. It had been assumed that pressure from overlaying rocks would prevent hot fluids from flowing through the cracks and pores to deposit minerals at such depths.

In the Dynamic Axis Theory we have discussed how crust quadrants are alternately subjected to tension and compression. In periods of crust tension, the cracks and pores deep within the crust also pull open to allow the leaching movement of hot fluids.

More surprises emerged when the drill core did not find the seismic indicated Conrad Discontinuity (assumed to be the border between granite-type rocks and the denser and deeper basaltic rocks at about 26,000 feet). This was an obvious temporary setback for existing theories on the meaning of seismic signals. The drill hole, now extended to 39,587 feet, has still not reached the basaltic rock. One of the premises for the Dynamic Axis Theory calls for deep-rooted intrusive magma to cool slowly and form igneous rocks with larger crystals. If this is true, basalts should not be found below the granites, unless the basalt layer that lays under the granite cooled rapidly during the earliest years of earth's life. Later the magma that froze to form the granite rocks would have had to pass up through the basalt basement, but remain below an insulating cover to cool slowly.

Another surprise came when the Germans drilled a pilot hole in the Oberpfalz area in preparation for a targeted 49,200 foot hole. The Oberpfálz area was selected because it was assumed to have a cooler crust. Being cooler, they did not expect to encounter the critical 300°C temperature level where subsurface rocks become ductile, before they reached their target depth. On the pilot hole, however, they reached 118°C at about 13,000 feet rather than the predicted 80°C. Revised estimates now call for the ductile region at about 32,800 feet. We are unable to project scientific understanding to our planet's interior, only a few thousand feet. This simply warns us to be very cautious about elevating a theory based on remotely sensed data to a working hypothesis to support other theories.

Coastal Flooding

Speculation about coastal flooding as a by-product of the greenhouse effect has received much attention in recent years. This is based on the assumption that worldwide warming will melt the existing ice sheets in Antarctica and Greenland. Several factors tend to counter the melting cycle. We have already discussed the role of heat in creating an ice sheet. Warmer ocean waters will lead to increased evaporation, thereby increasing the amount of water available in the hydrologic cycle. Increased heat also generates more volatile atmospheric currents. The collision of wet warm atmospheric fronts and cold fronts triggers violent weather with greater precipitation. Increased moisture in the clouds moving over Antarctica and Greenland could bring heavier winter snows that more than offset any summer melting by slightly higher temperatures.

Whenever tide gauge stations along any coast indicate a rising geoid, it may seem natural to attribute the change to the melting ice sheets with all the attention the greenhouse effect receives. The calculations of rising sea levels, however, need to take into account the effect isostatic balance responses have on the sea floor. If the additional ocean waters cause the sea floor to subside, sea levels may change very little.

As we have seen, both the geoid and the crust will respond vertically when the earth's axis drifts. We need to use highly precise VLBI geodetic measurements over an extended period to determine how much of any indicated change in sea level is attributed to the change in geoid and how much to changes in crust.

Lost City of Atlantis

The mythical stories about the lost city of Atlantis have provided fascinating speculation for many. Atlantis was a city of advanced technology for its time with gold and glitter to surpass the era of Egyptian pyramids. With much early history centered on the Middle East, the Mediterranean Sea has had its share of attention as the possible site for Atlantis. Probably many speculations, however, call for an area around the Azores. In any case, the assumptions are usually based on Atlantis sinking into the sea. Our discussion in chapter 7 presented evidence of vertical movements of the crust that could destroy a regional culture by drowning. With the Dynamic Axis Theory we have a simple explanation for a combination of sea level and crustal changes that could drain or flood any local on earth. It is not difficult to speculate, as suggested by Charles H. Hapgood's maps of the ancient sea kings, that an advanced civilization of long ago (maybe centered in Atlantis) could have known about the Americas and Antarctica.

The Deluge

To apply the Dynamic Axis Theory to the Deluge, we simply have to visualize a sudden shift of the earth's axis that set the oceans sloshing back and forth in their basins. Tsunamis, triggered by an axis surge and enhanced by seismic activity, could have drowned the area of advanced civilization. A by-product of an axis shift and the redistribution of the ocean basins would, undoubtedly generate very unsettled atmospheric conditions. Such conditions could trigger rains that lasted for 40 days and 40 nights. A sudden change in sea level could explain how Noah's Ark was picked up by the rising waters to drift aimlessly until it reached some distant land unaffected by the geoid shift. It could explain how ocean floor drained when the axis shifted. In the context of the Dynamic Axis Theory, the Deluge becomes a physical possibility.

Could the Deluge and Atlantis be based on the same geologic event?

9 THE SPECULATING GAME

Now that the Dynamic Axis Theory has been presented for consideration, we can speculate on just what happened thousands of years ago to bring on the ice age, and what caused the ice sheet to melt. Consider this as a brainstorming session where ideas are advanced without initially considering their merits. The idea is to ignore all preconceived notions, and to apply inverse the generally accepted logic of concepts. If today's logic holds that earthquakes cause the axis to shift, reverse this and suggest that axis shifts cause earthquakes. If we believe that ice ages are caused by periods of extreme cold, suggest that heat causes ice sheets to grow. In each case, take the alternate suggestion as far as you can. Using a why, where, when, and how question and answer session, see if you can find any support for the concept. In a sense, this brings me full circle since speculation about my grandfather's theory on the cause of ice age sparked a lifelong avocational study.

The Seed

How well did my grandfather's ice and coal age theory make out in this reassessment of the role axis movements play in our planet's evolution? Very well, thank you. All the ingredients for climate swings of the magnitude necessary to build and destroy continental ice sheets and tropical rain forests that he envisioned remain intact. Changes in latitude, elevation, and land-ocean distribution explained in chapter 2 were the foundation of my grandfather's original idea. Crustal response to the changes in the magma seas, provided by the expanded theory, only alters the regional magnitudes of elevation uplift and subsidence. Any regional uplifting of the crust in the positive quadrants will reduce the change in sea level. This suggests that larger axis shifts are needed to accomplish the same results as given in the theoretical one-degree axis shift scenario. Similarly, any regional subsidence of the crust in the negative quadrants will reduce the relative drop in sea level. New ice sheets and tropical rain forests can still be produced by axis shifts.

The Dynamic Axis Theory does not depend on global cooling and warming to account for swings between ice and coal ages. The same heat distribution furnished by today's sun will do the trick. Ocean waters evaporated and transported by warm air currents are deposited downwind. Temperatures as a function of latitude and elevation determine if the precipitation is snow that builds an ice sheet or rain that irrigates the tropical rain forest. Regions that receive reduced precipitation become frozen tundra lands or deserts.

Based on the books and articles available to him, my grandfather assumed that the axis shift eliminated the last ice cap about 20,000 years ago. Scientists, using carbon-14 dating, have since nearly cut in half their estimate for the peak of the Wisconsin ice cap advance I have been unable to reconstruct the timing he applied to his theory. It apparently came from his belief that each of the four regional ice cap advances represented an ice age, and that five climate periods had occurred over the past 100,000 years. Since we are near the end of the present 20,000 cycle (20,000 since the end of the last ice age using his available figures), he felt that the next axis shift could occur at any time.

My grandfather proposed that small axis shift could alter elevations and latitudes sufficiently to account for the formation of continental ice caps. He spent the rest of his life trying to identify the forces that might disrupt the earth's spin axis. One approach was to research books on astronomy. I remember a solar system mobile he built that took up a sizable part of his small living room. In particular, he searched for evidence of a large comet with a cyclic orbit. The orbit needed to bring it close enough to our planet every 20,000 years to trigger an axis shift as it passed. In discussing the comet theory, it was obvious that he had not convinced himself of its merits. I am sure he was unable to find any indication of a comet that would fit the bill.

Antarctica Ice Age and the Amazon/Congo Coal Age

The Ice Age and Coal Age create a vision of the cold and hot extremes of past geologic periods. However, we do not have to look beyond today's planet climate for examples that represent both ice age and coal age conditions. Scientists studying Antarctica's ice sheet have a laboratory equivalent to the continental glaciers that covered northern North America and western Europe. Likewise, scientists studying the vegetation of tropical rain forests of South America, Africa and southeast Asia witness the conditions that can produce the materials needed for future coal beds.

In my grandfather's mind, four glacial advances represented by the European, Labrador, Keewatin, and Cordilleran ice sheets constituted separate ice ages. He was aware that common lands were covered by ice sheets that spanned more than one ice age. He referred to the present age as the Greenland (or Antarctica) Ice Age, and felt the five ice ages represented a five step repeating cycle. Unfortunately, I have been unable to reconstruct his evidence and conclusions. I do remember my father placing four small squares of tape on a ten inch world globe to represent my grandfather's four additional axis positions. If we cupped our hand so that the tips of the four fingers and thumb formed a circle (about one and one-half inches in diameter), it was possible to touch each of the four squares of tape and the present North Pole.

Even if his five axis positions represent a repeating cycle on a circular path defined by axis shifts, it would not necessarily follow that the second cycle would exactly repeat the same series of ice ages. As the axis wobble, a drift of the center of the five repeating ice ages could eventually move far enough to account for ice sheets in equatorial regions and rain forests in the polar regions.

We have discussed the shifting weights that disrupt the spin axis. None would account for a repeating cycle of axis shifts. It would take something like the comet suggestion to produce a periodic axis shift. Paleomagnetic studies add strong evidence that the pole path is random, not a repeating cycle. Textbooks today treat all four of the above mentioned ice sheet advances in the Northern Hemisphere as one ice age, but with interglacial periods interrupting the major advances of the ice. The question remains, were the interglacial periods regional or global events? This is where dating techniques will be tested. The

Dynamic Axis Theory suggests that small axis surges are responsible for the regional climate shifts that produced ice sheet advances. If that is correct, scientist should be able to find indications that the timing of ice sheet decay in one region of North America or Europe matches the growth in another region.

The only written record of my grandfather's theory, a pamphlet printed in 1926 by William Chester Strain "The Mechanics of the Movement of the Equator" is reproduced in Appendix F.

Dynamic Axis Theory and the Ice Age

In most discussions about ice ages, people seem to operate on the hypothesis that global cooling was involved. This is simply based on the ice sheets found in cold climates. When evidence is presented about the existence of ice sheets in middle and equatorial latitudes, they reason that the global climate must have been colder. What happens, if only the temperature is changed? If the average winter temperature in northern Alaska and Siberia were lowered by 10 or even 20 degrees, would a glacial ice cap form? No, not necessarily; cold air alone does not produce snow and ice. Only if moisture evaporated in a warmer climate is transported into the region, will additional snow fall to form ice on the lands. If the average summer temperature of Greenland or Antarctica increase by 10 or 20 degrees, would the ice cap melt? Yes, or at least it would recede around the edges. Any land exposed when the ice sheet melts will absorb heat, thus increasing the potential for additional melting. If we overlook the role of heat in producing the required snow, we inadvertently limit our theory options. This is not to say that global temperatures have remained constant over geologic history—for surly they have not. The notions associated with the greenhouse discussions are based on known causative relations for heat trapped near the earth's surface. They do not by themselves, however, dictate global climate. The Dynamic Axis Theory introduces a long list of variables that must also be taken into account.

It is generally presumed that ice sheets melt faster than they form. Although this is probably true, we must be careful not to completely exclude the possibility of a rapid ice sheet buildup. Consider the scenario of an axis surge that placed warm ocean waters upwind from an elevated continent. Atmospheric currents feeding warm moist air over elevated lands where they encounter cold Arctic blasts could easily accumulate 100 feet of snowfall annually. If we assume that 90 feet of the annual snow melted during the summer and the rest compressed into one foot of ice, it would only take 5,280 years to create a mile thick ice cap.

One of the ice age theories that is presented in several forms, involves dust particles thrust into the upper atmosphere and stratosphere that blocked the sun's warming rays. The mechanics suggested for casting the dust into the atmosphere and stratosphere includes a rash of major volcanic eruptions, or a volley of asteroids striking the planet. Proponents cite "The Year Without Summer" associated with the eruption of Tambora in 1815 as proof that extensive volcanic activity can reduce global temperatures.

The decay rate for the debris from volcanic eruptions or an asteroid strike makes it difficult to imagine a significant, long-term contribution to an event that extended over a period of 10,000-20,000 years or more. The Dynamic Axis Theory accepts increased volcanic activity and asteroid strikes as only two of many elements for short term climate swings. The theory incorporates the idea that the frequency of volcanic activity is greatly accelerated by an axis surge. A simple test of this would be to compare the amount of volcanic debris in the first and subsequent deposits of each sedimentary bed. A higher volcanic ash concentration in the earliest deposits of each geologic age would attest that increased volcanic activity is associated with the global cataclysmic events. The latter is responsible for initiating dramatic geologic changes—possibly triggered by an axis surge.

In his search for an answer to ice age cooling, Milankovitch excluded the planet. He found very limited global fluctuations in total heat striking the earth as it follows its elliptical orbit around the sun. He noted cyclic variations in the axis tilt relative to the sun that could account for differential hemisphere cooling. When the earth axis tilt creates cooler summers, it preserves the winter snow fields. This also assumes that the milder winters do not significantly reduce the snow accumulation. In fact, warmer winters should make for heavier snows for the same reason that spring and early fall snows have more water content than winter snows. Since warmer summers occur in the opposite hemisphere, Milankovitch Cycles do not directly produce a global temperature swing. Some attempts to explain how Milankovitch Cycles might account for global cooling were mentioned in chapter 8. Long term fluctuations in the sun's energy out put are also a factor, but one that has not been quantified for geologic periods.

The Last Ice Age

Without more definitive data, we can only postulate on different axis positions that could have contributed to an ice sheet formation over Canada and western Europe. Wild speculations are more fun anyway. Any scenario we develop represents approximate guesses at best.

Although limiting ourselves to the role of the geoid's response to axis shifts, consider the following scenario. Once upon a time the earth's axis shifted, placing the North Pole several degrees south of its present position toward Greenland along today's 50 degrees west longitude. The regions now known as North America and western Europe had higher latitudes and higher elevations. Asia and the North Pacific Ocean had lower latitudes and elevations. In the Southern Hemisphere, Australia and Wilkes Land Antarctica had higher latitudes and elevations, while South America, Antarctica Peninsula, and Africa were lower. As in most discussions about the Ice Age, we will focus our attention on the ice sheets of the Northern Hemisphere.

Expanding the original theory to include the earth's crust response to variations in the magma sea pressure introduces many unknowns into the equation. The theory of plate tectonics suggests that North America and Europe were closer (back up the ocean spreading in the Atlantic to the beginning of the ice age). If the variations in magma pressure were absorbed by raising and lowering the thinner crust of the ocean basins, most continental changes could be calculated in the same way as the hypothetical 1-degree axis shift. An axis shift of just over 4 degrees could make the Canadian Shield a mile-high plateau.

The resulting higher geoid in the Bering Straight region allowed the warm Pacific currents to circulate northward into the Arctic seas. There they exchanged more warm and cold waters between the primary circulation eddies of the two oceans.

The warmer Arctic Ocean increased the moisture content of the clouds that were carried by prevailing air currents down across an elevated Canada. Canada's higher elevations would provide cooler temperatures and trigger the precipitation as snow. Once the highly reflective snow had blanketed the land, solar heat was reflected back into space, thus holding down the temperatures even during summer months. Cloud cover during the summer months also reflected solar heat back into space and reduced the amount of melting.

When the jet stream shifted to bring in the warmer Pacific waters, their moisture laden clouds moved in over the western United States. They filled Lake Bonneville and the other depressions of the Great Basin to their brim with rain water and snow melt. Moisture laden clouds, reinforced by evaporation from the western lakes, moved to the Northeast where they collided with cooler air from the north. Several heavy snows blanketed the Canadian Shield region each fall, winter and spring. Since more snows fell each year than melted during the summer, an ever thickening ice sheet formed.

Similarly, Europe's moisture came from the middle latitude of the North Atlantic. The narrower Atlantic Ocean made its circulating pattern more elliptical. Ocean waters remained warmer because of the shortened path; specially if the flood basalt eruption in the Iceland region served as a hot plate to add power to the ocean humidifier system. Prevailing atmospheric currents feed the moisture laden clouds over an elevated western Europe to deposit as snow and create an ice sheet. Remember, the fjords of Norway indicate Europe could have been nearly a mile higher during the Ice Age. If valley glaciers can carve fjords below sea level, scientists should be available to find examples along the coasts of Antarctica and Greenland. The Ross Ice Shelf and Ronne Ice Shelf in Antarctica, however, demonstrate that glaciers feeding into the ocean soon lose their sculpturing force. Add a touch of Milankovitch cycle, a rash of volcanic eruptions thrusting dust to the atmosphere and stratosphere, and one of Nature's swings in greenhouse gasses and we have the recipe for creating an ice sheet. Did I not tell you that speculating is fun? The hard part comes when scientists try to find solid clues for the existence of anyone or all the above variables.

Similar scenarios could be developed to speculate on other climate extremes. Did an increase in moisture during the ice age add vegetation to the Sahara, and develop its now buried drainage system? Was some other region, presently covered with vegetation, converted to a desert in the shadow of moist air pattern?

Alternative scenario

An axis position south of the present pole along about 110 degrees west longitude would provide for higher latitudes and elevations for North America, but not Europe. Here we would need to call on the higher magma seas to elevate Europe sufficiently to trigger the snowfall if the two continents' ice sheets occurred at the same time. The interesting aspect here is that an axis shift to the present pole would place the entire Mid Atlantic Ridge (an S-shaped alignment of midocean ridge) under the stretching influence of positive quadrants. The same shift would have created negative quadrant compression on the crust for the series of deep ocean trenches (subduction zones). These ring the western Pacific in an S-shaped alignment from the Aleutian Trench to the Kermadec Trench north of New Zealand.

Dynamic Axis Theory and Coal Ages

The Amazon, Congo, and southeastern Asia jungles are cited as the type of vegetation needed to create coal deposits. Yes, rain forests and ice sheets coexist on the planet today. Rain forests represent an extreme of a moist and usually warmer climate on the planet. Cold and hot arid regions also coexist in places such as Siberia and the Sahara. Although global temperatures undoubtedly fluctuate over geologic ages, global cooling and warming is obviously not required to account for the evidence of massive ice sheets and coal deposits.

For coal ages we need geologic periods that provide a massive supply of decomposed vegetation. A sizable ice sheet can be created and destroyed in 10,000 or even 100,000 years. The rain forests, assumed to be the source of coal deposits, could certainly have flourished and died over similar time spans, assuming appropriate climate conditions existed in a region.

The Dynamic Axis Theory adds a new dimension to the speculation about coal bed formations. Do coal beds represent the sites of natural growth and decay of a lush rain forest that fall in place and are later buried? In time, will heat and pressure decompose the vegetation to form peat, bituminous coal, lignite coal, or eventually diamonds? Consider an alternative scenario. An axis surge creates massive tsunamis to sweep across the lands and sets adrift masses of flora and fauna stripped from the lands. In this scenario, it would not be necessary for the vegetation to come from lush rain forest lands. All the vegetation debris stripped from Kansas and Nebraska is set adrift on the oceans. It came to rest in inland lakes, marshes, coves, or the shallow bay waters along the new coast lines and created a sizable brush pile. Thick beds of vegetation deposited in the shallow waters and overlaid with sediments would begin the decay process that evolves into coal.

The formation of 200 feet thick coal beds in the Powder River Basin of northeastern Wyoming and southeastern Montana leaves scientists puzzled. Most coal seams are much thinner, and have influxes of sediments that separate the coal seams. Since the Powder River Basin coal is not interlaced sediment seams, the fuel for the full 200 foot thick bed must have accumulated as a unit. For the vegetation to turn to peat and later coal, required pressure and heat. Pressure calls for a thick sedimentary cap. Heat conducted up from the planet's interior afforded the cooking. Now the Powder River Basin coal seam is close enough to the surface to make strip mining economical. That means that most of the thick overburden of sediments has been eroded away. As the Powder River Basin coal is low in sulfur also shows that the vegetation was not deposited in ocean waters that contain sulfates.

Consider what would happen if the tidal waves swept all the vegetation from the northern Appalachians and deposited it in Lake Ontario. Lake Ontario is an area about the same size as the Powder River Basin. The next step would be to seal the debris with a layer of sediment. A second axis shift pushed the region below sea level on the continental shelf. Here thick sediment layers added pressure to convert the vegetation to peat and coal. This scenario is presented only to illustrate how large thick coal beds could be formed without tropical rain forests.

If we repeat this process of tidal waves sweeping the lands several times, but on a smaller scale, we would answer the question of how multiple coal beds separated by layers of silt, sand, and gravel could have been formed.

Dynamic Axis Theory and Regional Geology

My grandfather's theory called for higher latitudes and elevations for the North America and western Europe during the last cycle of Ice Ages. A higher elevation for the continents can account for some of the lower sea levels (drowned marine terraces). This is usually attributed to more ocean waters being trapped in the ice caps. On the other hand, the isostatic balance theory suggests we should find scars of higher sea levels when the continent was pushed deeper into the magma seas. This, of course, assumes that the ocean floor also responds to isostatic balance forces to maintain a near constant sea level datum. The uncertainty presented by these opposing forces points out the critical need for dating vertical crustal movements to try to establish the sequence by computer modeling.

Mississippi River Basin

When the ice sheet moved across the St. Lawrence River Valley into New England, the drainage outlet to the Atlantic was blocked. Consequently, drainage from summer melts during the height of the ice sheet sent silt laden waters down the Mississippi River. During the ice sheet decay, the volume of the Mississippi drainage basin would have exceeded many fold any of today's rivers. In time, the glacier that blocked the St. Lawrence drainage melted, and the flow down the Mississippi was reduced.

If the North American continent was higher, say by 500 or 1,000 feet, the Mississippi Delta would have been far out in the present Gulf of Mexico. The gradation of sediment deposition, from coarse to ultra fine, would shift southward along with the Delta. The sediment type would reflect the outwash from the glacier. Although erosion from the eastern slope of the Rockies was part of the mix, the source material would have been considerably different from what is presently being deposited at the Mississippi Delta. Such changes in source material and mechanics of deposition can explain breaks in the sedimentary bedding around the world—clues that can help reconstruct geologic history.

The higher continent elevation would accelerate erosion. This includes the carving of a canyon near today's Mississippi Delta that has since been filled with sediments. Analyzing cores from wells drilled along a line perpendicular to the Mississippi River Basin should provide information about the possibility of a buried canyon.

My grandfather's theory called for a sudden axis surge that brought the geoid closer to its present levels and initiated the rapid ice sheet decay. The meltdown had to be triggered by some combination of imported warmer air currents during the summer, fewer below freezing days during the winter, snow-melting, dry winds, and heavy summer rains. The ice sheet decay, in itself, constituted a sizable weight shift that would call for additional axis adjustments by drifting or surges.

Flip Side of the Mississippi River Basin

Imagine a much higher geoid that floods the lower Mississippi as far north as Cairo, Illinois. The present Mississippi Delta would be far out in the expanded Gulf of Mexico, and receive only a very thin layer of ooze and ultra fine silts that remained in suspension. In addition, wind blown dust, volcanic ash, and meteoric dust deposited directly on the ocean waters. The new Mississippi River Delta, above Cairo would receive only the sediments from the upper Mississippi River. The Ohio, Missouri, Tennessee, Arkansas and Red Rivers would all have their deltas. Each river delta would receive sediments characteristic of the individual river basins. The rate of accumulation would be much less at each of the individual river deltas than the present combined accumulation in the Gulf of Mexico.

The above scenario serves only to visualize how changes in the geoid can account for breaks in sedimentary bedding. If, at some point in geologic history, the geoid did intercept the Mississippi River and its tributaries as suggested, the uplands would have been far different from what they are today. Rivers from the west would have drained the ancestral Rockies. How large were they? To the north, was the Canadian Shield a high plateau region of sedimentary beds feeding silt laden streams? Did mountains even exist to the east (pre Appalachians)?

America's Mediterranean Sea

Were there deep, sediment filled basins that lined up from west Texas through Oklahoma, Arkansas, Kentucky, and into, Michigan North America's Mediterranean Sea? The area involved is somewhat smaller, but the depth of the basins exceeds the Mediterranean Sea. Sediments have since filled these depressions to leave no surface indication of their existence. Now that they are well above sea level, their soils are being eroded and carried off to the Gulf of Mexico. If the Canadian Shield was the basement rock supporting a massive mountain range or thick plateau sediments, they must have been much larger than the present Himalayas. The above mentioned basins had to be filled before sediments from the Canadian Shield region contributed to the filling of a much deeper trench that existed along the coast of Texas, Louisiana, and Georgia.

Expanded Dynamic Axis Theory

The expanded Dynamic Axis Theory offers many more options to the ice age theory than perceived by my grandfather. I do not remember him trying to correlate it with mountain buildings, mass extinctions, earthquakes, or volcanoes. Judging by his enthusiasm in explaining his thoughts, it would have given him an exciting challenge, if he had the opportunity to contemplate scientific understanding from today's base.

10 WHERE DO WE GO FROM HERE?

Trying to keep track of all the players in the earth science arena can be compared to watching a football game. By design, all 22 players on the football field have a specific assignment to carry out in each play that will last only a few seconds. Thanks to multiple cameras, instant replay, and slow motion, the coaches can find many things to criticize after the game that even they had missed. Each team can select from many offensive and defensive plays along with several options for each alignment. Even more options are introduced when one or more players fails in their assignment and begins to improvise. When the ball is snapped, things seldom happen exactly as the coach has charted in the play book since it is the defense's objective to disrupt all designed offensive plays. The multitude of variables met throughout the game generates the excitement in the avid fan. If we concentrate on one player, such as the quarterback, when watching a football game, we will miss most of the other 21 activities. We also forego the actions of the officials, coaches, others on the sideline, and even people in the stands that may influence the outcome of the game. Only by developing peripheral vision can we come close to seeing the game from the coach's perspective.

We are all fans of the game of earth science. The objective of the first nine chapters has been to present the Dynamic Axis Theory. It is one of many paragraphs in the earth science play book, and intends to add some degree of understanding to Nature's mix of orderly evolution and chaos. Nature's game plan is much more complex than a simple football game. The challenge, however, is the same as we attempt to learn the rules and observe the game as it unfolds. Regardless of which contributor to an earth science event we concentrate on, we must attempt to remember how the other contributors are affecting the outcome. When we zoom in on a micro-perspective of chemistry, biology, or other scientific specialization, everything observed must also be considered from the macro-perspective. When we step back to look at the big picture, we cannot completely ignore the minute detail.

Put the Dynamic Axis Theory to the Test

When a theory is advanced, it becomes fair game for any challenging tests that can be conceived. Each time a test of the theory comes through with a passing grade it adds support. When a test falls short of expectations, it is back to the drawing board for a reevaluation. In some cases a simple fine tuning of the theory is all that is needed. More serious deficiencies call for more drastic remodeling of the theory or segments of the theory. The theory is a complete failure only when none of its new concepts can stand up to the test. Undoubtedly, there are experts with many of the answers to the questions raised in the preceding and following paragraphs. The problem is to pool the existing knowledge of earth science and address the questions raised from the perspective of the Dynamic Axis Theory. You are now a member of the jury that will deal with the evaluation.

Data Base

Fortunately, we have volumes of earth science data that were collected over the years. Unfortunately, much of it is presented in a different reference system making it difficult and sometimes impractical to employ much of the acquired information in computer models. Since the Dynamic Axis Theory involves global events, and the data volume requires computer assistance, all spatial relationships need to be presented in a single geodetic reference system. It is never too soon to think about designing and adopting a universal data base that provides an inventory of the physical features of our planet.

A universal data base needs to accommodate: (i) point features (geodetic control monuments, earthquake epicenters, location of fossil discoveries, and instantaneous pole positions); (ii) linear features (the trace of marine terraces, ancestral streams, surface scars of fault lines, and plate boundaries); (iii) area features (ancestral seas, flood plains, climate regions, land and water bodies); (iv) surface data (fault surfaces, upper and lower surfaces of each sedimentary bed, Moho, Digital Terrain Models [DTM's] that define the earth's crustal surface, using a grid array of elevations); and (v) volume features (sedimentary beds, ice caps, and ancestral mountains defined by upper and lower digital surface models and their area boundaries). Each feature has from one to several attributes that provide an additional subdivision of the feature data (fossil type, well log, stream flow volume, vegetation type, sedimentary deposition mechanism, chemical analysis, earthquake intensity, etc.). Features may also have text information in narrative form that help the data base user, but does not fit into the designed inventory format. The data base includes all the features on topographic, geologic, meteorological, and other earth science maps that represent an inventory of our planet.

Each feature needs to be defined by some spatial coordinates. Several geodetic reference system options are available for the feature address system, including (i) any one of many conventional (geographic) latitude, longitude, and elevation coordinates reference systems; (ii) the military's Universal Transverse Mercator (UTM), using Cartesian X, Y, and elevation coordinates in 6-degree zones of longitude; or (iii) a geocentric coordinate system with Cartesian X,Y,Z coordinates.

Both geographic and Universal Transverse Mercator coordinate systems provide easily recognizable addresses that computer buffs would call "user friendly." The Cartesian coordinates of Universal Transverse Mercator provide a natural cell array for raster scanning and mating. The geocentric reference system, with its Cartesian coordinates, provides the simplest data organization, making it computer friendly, but not user friendly. Since a geocentric coordinate system is an awkward system for collection and visual applications of earth science data, the computer programmer may choose to employ it only for data transformation and computation within the computer program. Regardless of the computers internal program design, both input and output have to remain in some user friendly geodetic reference system.

When attempting to reconstruct geologic history, the sequence of events must be known. The conditions before and after a geologic event may provide the clues why the volcano erupted, the crust ruptured, or why sea levels changed. Consequently, two attributes that need special treatment are "time" and "accuracy." The dynamics of earth change dictate that "time" needs to be treated as a critical fourth dimension. Economics, likewise, dictates that data "accuracy" needs to be included as the fifth dimension. Not all data can or should be collected with the same precision. For example, it is impossible to define the Sahara's boundary or the sediment volume deposited by the Mississippi River during the Ice Age to the same accuracy as is needed to define the boundary of Lake Michigan or the volume of rock blown off the top of Mt. St. Helens during its eruption. Time units also vary in precision. Units for an appropriate rate of change will vary from seconds for earthquake data to thousands or millions of years for defining geologic ages. The positional accuracy of geodetic control points has the greatest demands on both absolute and relative computational precision to provide a datum to detect tectonic plate movements. On the other end of the scale are the boundary lines or limits of desert regions, ancestral ice caps, and climate zones. In between are features or activities such as earthquake epicenters, crustal faults, drainage basins, rivers, and lakes. Evolutionary change can be interpolated between selected fixed time points. Cataclysmic surges become critical time points that alter or distort evolutionary change.

The so-called string data is one method of digitally recording linear features and area boundaries (or any linear feature). This is similar to the way children's dot to dot pictures are defined. For computer modeling, data organized in arrays that represent a natural subdivision of the planet offer many advantages. Organizing the data in arrays of cells or pixels allows earth scientists to manipulate data in the same way as a television picture that is sliced into its primary colors for transmission, and then reassembled on the home TV in its original form. Computer modeling programs can apply testing algorithms on a cell by cell basis or by any statistical grouping when lower precision is sufficient.

Cell nesting enables us to accommodate the variations in resolution requirements. For example, climate data, such as average annual rainfall may only justify 10,000 x 10,000 meter cells. A finer resolution may require 10 x 10 meter cells or even 1 x 1 meter cells to define streams and ponds. Geodetic controls for the detection of tectonic plate movements and earthquake stress require sub-meter measurements. All point features can be recorded and stored as lists of Cartesian coordinates and easily merged with array data during computer manipulation and modeling. As long as the larger cells are even multiples of the smaller ones, computers can easily generate the necessary associations between data sets at any desired resolution.

Data compression is needed to reduce the tremendous array data volume. Assume that all features are to be collected at the finest resolution (say 1 x 1 meter cells). Data compression can be accomplished by nesting, using the largest cell that will preserve the integrity of the data. For example: to define Lake Michigan, 10,000 x 10,000 meter cells define most of the open waters in the lake center. Around the perimeter 1,000, 100, and 10 meter cells can fill in the finer detail of the shore line and bays. The computer can easily expand the single 10,000 x 10,000 meter cell into 100,000,000 1 x 1 meter cells for computation purposes if needed.

The objective of a universal data base is to provide a place for everything and to keep everything in its place. As earth scientists begin their inventory, an entry needs to be made in the data base indicating whether each feature has been inventoried. As the process continues, the inventory data replace the no inventory statement on an area by area basis. For the user it may be as critical to ascertain the absence of a feature in an area as it is to know an existing feature. For example, defining ancestral seas will always have many gaps in the data, but the knowledge of how complete the data set is establishes the confidence level of any derivative study.

Computer Modeling

Computer modeling programs offer the best hope for analyzing the mass of data that comes into play on each geologic event. Having the data in a universal data base will facilitate preparation and execution of computer modeling programs. For data collected and stored in a geodetic reference system other than the one used for the universal data base, conversion programs are needed to access the data.

Mechanics of Axis Shift

Before discussing computer modeling programs for testing the Dynamic Axis Theory, it may be desirable to use space technology to confirm the axis shift mechanics. To test the notion that weight shifts on the spinning orb are responsible for axis wobble, axis drift, and axis surge, consider conducting the following experiment in the weight-free confines of a space station:

Equipment. Design a solid globe that represents a miniaturized earth, and equip it with known weights that can be shifted on and within the spinning orb by remote control. Paint longitude and latitude lines on the globe as references that can be photographed by high speed cameras during spin tests. Include a post at each pole to be used to observe axis wobble and to establish the initial spinning of the globe.

Experiment. Set the dynamically balanced globe spinning true on its axis in the weight-free environment of a space station. Use remote control switches to slowly shift known weights on or within the globe. Photograph the spinning orb with a rigidly mounted high speed camera. Also record a synchronized record of each weight shift so that follow-up cause and effect relationships can be made.

Evaluation. Use the photographs to observe and measure any wobble of the globe's axis resulting from the shifting weights. A wobble will be reflected in both the poles and the longitude lines painted on the globe. Compare the wobble amplitude with the eccentricity of weights on the globe. Use the photographs to measure any drift of the globe's spin axis (axis shift relative to the globe mass). Establish if the axis shift occurs smoothly or in periodic surges as axis wobble reaches a critical point. Compare the results of the above experiment to the theoretical responses computed using the fundamental laws of dynamics.

Monitoring the Earth's axis

Another test of the earth axis' sensitivity to weight shifts uses VLBI technology to monitor the axis wobble at hourly intervals to check for a possible correlation with lunar and solar tides. Seek evidence for a time lag between weight shifts and the axis response. In addition, any anomalies in the axis wobble path attributable to large volcanic eruptions, spring runoff, or other known short term weight shifts, would lend support to the Dynamic Axis Theory. A major reevaluation of the Dynamic Axis Theory would be required, if any proof emerges that weight shift-induced stresses do not accumulate to trigger axis surges.

Computer modeling axis shifts

Computer modeling is valuable in analyzing axis response to known (or estimated) physical weight shifts on or within our planet. Use formulas derived from dynamic theories and the tests discussed above for a computer modeling program that calculates the earth axis wobble, axis drift, and axis surges as a function of weight shifts. Develop an inverse computation to determine the size of a force couple above the axis response to known weight shifts that is required to account for the observed axis wobble and axis drift.

Compute the axis wobble and drift. Assume the spinning planet at the beginning of the period is in dynamic balance. It is a function of sediments transported by the global river system and occurs over 14 years (ten wobble cycles) and 140 years (100 cycles). Compare the calculated axis wobble and drift with those observed since 1900. Use the difference between these values in a reverse calculation to find any unknown force couple. At least part of any missing force couple may be attributed to weight shift in the magma sea currents. Remember, in computing the axis response to weight shifts, it is necessary to include the geoid and crustal response to an axis shift as an additional weight shift.

Using a reverse computation, compute the axis drift and axis surge associated with the decay of the North American and western European ice caps. Assume the ice caps formed over the same span of time. Include the weight shift for evaporation from the Great Basin lakes. The objective of this calculation is to determine (in reverse) a theoretical pole movement over the decay period as a function of the ice sheet melt down. Several other hypothetical weight shift computations should be designed to establish a pattern for the global response to known and suspected changes in its physical loading.

Sediment Sources

A very large computer modeling program and many new data are required to establish the continental areas and mountain ranges that surrendered their gravel, sand, soil, and silt to form the world's massive sedimentary beds. For each bed one needs to determine the mechanics and direction of deposition flow. Hundreds of vectors are also required for the deposition flow patterns. Reversing the deposition vectors provides a clue as they point upstream toward the sediment source. One has to consider the possibility that the source may be a continental land mass now submerged under ocean waters or not presently attached to the continent. It is of special interest if many sedimentary bed vectors indicate the Canadian Shield region as their source.

Computer modeling of sediment deposition mechanics has to be applied independently to each sedimentary layer. Major shifts in sediment sources will reveal how the geoid and the crust responded to an axis shift. They also define the land-ocean patterns for each period of geologic history. Unfortunately, it is impossible to obtain a complete record of sediment mechanics data for the deeper sedimentary beds to insure high integrity.

Geologists use grain size, shape, and uniformity as clues to distance traveled, and method of transport to supplement the sediment mechanics from grading deposition. Chemical composition may also provide clues as to sediment sources.

Past Geoids

We also need a massive computer modeling program to reconstruct geoids of the past from marine terraces (ancestral ocean beaches). As with all detective work, the clues seem to diminish at an exponential rate with increasing time as one tries to unravel historical events. More ancestral marine terraces are eroded or buried under new sedimentary deposits than are exposed on today's earth surface. The task of reconstructing ancestral geoids depends primarily on accurately dating marine terraces that define sea level at some point in geologic history. The next step is to mathematically make a geoid best fit to the marine terrace evidence for each geological age. If the crust was rigid, as assumed in chapter 2, the computation is relatively simple. Its accuracy depends on the distribution of marine terraces around the world. Extensive records for one age distributed around North America would be much weaker than one or two records on each continent. With crustal gymnastics introduced in chapter 3, the task of determining a best fit geoid becomes extremely complex. Clues to local crustal warping are deductible from the sedimentary bedding. Less obvious are clues to regional uplifting and subsidence from increased and decreased magma pressure acting on the crust.

We are not limited to signatures from sea level scars. The plane created by large perennial lakes, such as Lake Bonneville terraces, can be considered parallel to the geoid for its age in geologic history, even though they are not at sea level. The sedimentary layers in large lakes or continental shelves offer other clues to ancestral geoids. In any case, it will be necessary to determine how much plane deviation is attributable to either geoid rotation or crustal gymnastics.

Marine terraces associated with the geologic age from the peak of the North American and European ice caps to the present should provide many answers. Since they are relatively recent events, fewer will have been destroyed. For every identified fluctuation in sea level, it is necessary to search for an opposite sea level change in the adjacent quadrants, and a companion fluctuation in the opposite quadrant. Of course, it will be necessary to eliminate the influence of isostatic balance adjustments. This specially applies around the glacier perimeter covered continent and at the delta of major rivers. The best signatures for the Ice Age sea levels may be found in eastern Asia, Australia, Africa, and South America, where there is a better chance to avoid the influence of ice sheet weight.

Another series of geoid computations that could clear up geologic history would be the theoretical geoids associated with pole positions plotted from paleomagnetic data. It will be necessary to make some educated guesses how the crust responds to each of these axis shifts. One possibility is to assume the ocean crust flexes to absorb most of the shift in magma pressure, thus leaving the calculated geoid changes essentially unchanged except for local magma intrusions (mountain building). When that does not work (and it probably will often not) other guesses must be considered. Uncertainties in the changing patterns of continents complicate geoid studies of the distant past, as the tectonic plates are propelled around the globe.

When researching geoid changes resulting from axis movements, scientists must be aware of the change in equatorial bulge resulting from a reduction in the earth's spin rate. Field geologists need to search for clues that directly address this concern. Computer modeling programs must also consider this additional variable.

Crustal Gymnastics

The Dynamic Axis Theory calls for magma intrusions into and through the crust under tension, primarily in the positive quadrants. We should find in the negative quadrants major mountain building episodes involving anticlines, synclines, monoclines, and overthrust belts associated with compressive crustal events for the same geologic periods. An exception to the rule is volcanic activity triggered by increased magma pressure. The latter arise from the hydraulic plunger action of subduction plates being driven into the magma seas in negative quadrants. Computer modeling programs have to address crustal gymnastics around the globe. Again, a critical element for reconstructing geologic history is to establish a starting and ending date for each event.

A computer modeling program has to determine the crustal extensions and shortening as a function of the higher and lower magma seas created by the Chandler wobble and axis drift. Contributing elements are crustal spreading along the midocean ridges, shortening along new mountain ranges, and crust lost to subduction. The transverse fault zones associated with the midocean ridges and relative movements between each adjacent tectonic plate provide additional clues of the tension, compression, and shear stresses. The Dynamic Axis Theory attributes these to the crust responding to changes in the oblate ellipsoidal shape of the magma seas.

A correlation between seismic activity and the earth's pulses could serve to confirm or challenge the hypothesis that the crust responds to Chandler and gravitational cycles. The same program can be used to analyze changes in crustal tension and compression associated with the axis drift since 1900. Today's axis drift calls for increased crustal tension in Asia and South America, with increased compression on the crust in North America and Australia. Additional clues to the role of pole movements may be discovered by using the program to analyze crustal tension and compression since the end of the Ice Age. We can assume either the present rate of drift or the pole position derived from ancestral geoid findings. Going even further back into geologic history, the computer program can analyze axis movements indicated by paleomagnetic studies. One needs a correlation between computed positive and negative quadrant episodes and physical evidence available for each geologic age insofar as is possible.

A computer program has to also predict the forces driving tectonic plates as a function of axis drift or axis surges. Another program should indicate what pole positions account for the spreading, subduction, and shearing zones between today's tectonic plates. Axis movements need to account for offsets in the transverse faults along the midocean ridges, slippage along plate edges, plate spreading, and plate subduction.

A computer program should analyze the forces acting on each segment along the San Andreas Fault, using the forces acting on the North American and Pacific plates. Remember that a plate's movement is not necessarily the same direction as the forces driving the plate. This program has to determine friction locks for each fault segment as a function of the acting forces, their dip and strike angles (including curves in the fault zone alignment), as well as the estimated rigidity and roughness of the fault surface.

Some clues may be derived by working the computations backward for areas such as Hollister, California, where stress is released by fault creep and small surges and indicates a weak friction lock.

It is desirable to continue monitoring the earthquake activity as a function of atmospheric loading as Jerome Namis suggested. SEASAT type data can be used to check the isostatic balance hypothesis as it applies to high pressure atmospheric cells displacing an equivalent amount of ocean waters. Low pressure tropical storms add an equivalent amount of ocean waters to maintain a near total weight balance.

It is necessary to determine which part of each mountain building episode is a product of crustal tension (magma intrusions) and crustal compression (folding and overthrust distortions). Evidence of tension events should correlate with compression events in complementary quadrants. The effect of the earth's slowing spin rate adds another variable to the crustal gymnastics equation, as it did the geoid equation.

We have to monitor VLBI and GPS stations for a possible correlation of crustal massaging by earth tides to establish their role in modeling our planet. A further correlation should be checked between earth tides and seismic activity.

Climate

First, we have to determine how much ice age climates are attributable to global temperature swings. If it is proven that ice and coal ages are a product of global cooling and warming, the Dynamic Axis Theory would suffer a severe blow. On the other hand, if evidence of past cold regions is related with warm regions, the theory would stand tall.

The Dynamic Axis Theory directs scientists to critical locations in search of evidence. Ice sheets in North America and western Europe suggest we should find signs of companion warmer climates in Siberia and South America. For example, reports from Siberia indicate that plants and animals belonging to a warmer climate existed before the mammoths were quickly frozen. As mentioned earlier, if the switch toward a colder climate in Siberia can be correlated with the warming of North America and western Europe, the Dynamic Axis Theory would receive a large boost.

Evidence of valley glaciers in the Andes is no proof of global cooling. Remember that heat generates heavy snows. Heavy snows combined with elevation can cause snow fields that survive the summer, even at low latitudes. More direct evidence of quadrant climates may be obtained from a study of glacial conditions on opposite sides of Antarctica during the last Ice Age. An axis shift, as suggested by the Dynamic Axis Theory, should extend the snow cap northward on one side of the continent, whilst it decays on the opposite side. The theory calls for a colder climate on the Australian side of Antarctica and a relatively warmer climate on its Atlantic coast during the last Ice Age. Evidence of cold ocean currents or high oxygen-18 to oxygen-16 isotope ratios in today's warmer ocean currents must be evaluated with complementary test sites. A special computer modeling program has to address variables, such as variations in deep and surface ocean currents, fresh water patterns entering the ocean basins, and depths of test sites at the time of deposition. All these elements affect the oxygen isotope ratio. Accurate dating of each environmental clue is essential to resolving the puzzle of our past climate.

Climate modeling programs are a vital part of any Dynamic Axis Theory testing. Regional climate change is a natural product of movements that alter the atmosphere and hydrosphere. The computer modeling programs developed to study the possibility of a greenhouse effect include many of the features needed. The programs have to be modified to accommodate different land-ocean configurations, land elevations, ocean basin depths, atmospheric and ocean current patterns, and latitudes for the hypothetical axis shifts suggested by the Dynamic Axis Theory.

As pointed out earlier, extreme attention to detail has to used in dating events that can change an ice sheet to an ice-free continent in a few thousand years. Glacier scars in North America and western Europe cannot be taken as a worldwide event without further proof. Any phase differences in the interglacial periods on the two continents could provide valuable clues to regional as opposed to global or hemisphere climate swings.

Some of the requisite input involving crustal gymnastics derives from the geoid computer modeling program noted above. One of the objectives is to determine global atmospheric temperature and precipitation patterns. The computed distribution of potential areas of glaciers, tropical rain forests, mid-latitude forests and grasslands, hot arid deserts, and arid tundra lands has to be compared to ground truth to determine if the program weighing factors are reasonable.

Coal Ages

The study of coal ages may offer a better opportunity of delving into climates of the distant past. Evidence of coal ages is easily recognizable throughout the world since the signatures are locked up in the rocks. Multiple coal seams separated by sediments provide a similar story as interglacial periods. Coal seams furnish more geologic history signatures in the form of fossils. The seams locked between other sedimentary layers provides them with a much better chance of survival than the surface scars of glaciers. Dating of coal seams should be much more accurate than dating ice age scars. Of course, the glaciers' moraines and outwash will provide some evidence, but both are very vulnerable to future erosion that destroys or smears the signatures.

If it can be established that coal and ice ages coexisted in the past as they do today, the identification of coal deposits will determine where to look for the corresponding ice caps, or at least colder regions. Continental drift will cloud some of these records.

Although concentrating on evidence of coal deposits and glaciers, we should not overlook the indicators of dry climates, namely: burning desert sands and arid frozen tundra. Again their global locations are needed to complete the atmospheric and hydrologic patterns for each period in geologic history.

The sudden climate swings indicated by the Dynamic Axis Theory are critical signatures for testing. Areas subjected to the widest swings are associated with the shift meridians. Simultaneously, the neutral meridian and neutral equatorial great circles should experience much less climate change. Global atmospheric current patterns could, however, interchange arid and wet regions. The global patterns of large climate change combined with areas of little or no change in themselves become critical signatures of past geologic history.

Mass Extinctions

One side effect suggested by axis surges is that mass extinctions are more often regional than global events. Regional and worldwide extinction patterns correlated with climate and physical changes in the geoid should fit predictable patterns within the overall Dynamic Axis Theory. Of particular interest is the ongoing debate over a sudden as opposed to gradual demise of the dinosaur.

Two specific signatures should stand out in the search for evidence of mass extinction. One is the mass grave of animal species that attest to a catastrophic event. Historians have pieced together how ancestral American Indians hunted buffalo by studying the remains found at the kill sites. However, evidence that is vulnerable to the erosion process or decomposition will not remain for millions of years, unless it is buried and preserved as fossil evidence. The second signature comes from dating the most recent evidence of a lost species existence. Of course, if the latter is in a mass grave (higher than normal concentration of fossils), we can assume that the physical cataclysmic event eliminated or weakened the species. They could not survive the subsequent climate change and disease epidemics. The distribution of the most recent lost species becomes an important signature for determining the method of mass extinction.

Sedimentary Bedding

Most geologic evidence comes in one form or another from the sedimentary beds. The bulk of the information for climate and the evolution of life forms is found here. There is a marked difference in the deposits laid down today at the delta of the Amazon River and that of the Yukon River. Deposits from each river basin have their unique characteristics. Today's fossils to be found by future geologists will vary from region to region. Dating techniques must be improved before much progress is possible in computer modeling of sedimentary beds around the world. Otherwise scientists will be mismatching earth science data and draw incorrect conclusions.

Predictions

We have briefly mentioned the problems associated with predicting earthquakes. This is considered a child's play compared to predicting axis surges. We do not even have one surge event during recorded history to serve as a model. For now, any attempt to place time restraints on axis movements is incomprehensible. The temptation to try to predict future axis surges must be weighed against the potential damage it would cause in fear and apprehension. This applies, even after we reconstruct past geologic history from the perspective of the Dynamic Axis Theory. There will be a foundation for a prediction of the next axis surge only if some predictable cycle becomes unmistakably obvious.

You are the Judge

We can only go so far in making direct observations about many earth science events. After many repeated observations, scientists are able to establish some natural laws that allow us to predict what to expect under controlled conditions. Not all natural events lend themselves to direct observation. When this occurs, scientists deduce a logical scenario of events based on the facts as they are perceived—a theory is born. The acceptance of a theory depends on how few doubts remain. Acceptance, however, never guarantees substance.

The Dynamic Axis Theory is composed of many concepts (sub-theories). They can be evaluated by rating each idea on a scale of one to ten, with one being impossible and ten highly probable. Many of concepts are listed below along with other theories that address the same topic. The "other" theory list is obviously incomplete, and can be supplemented with any additional theoretical explanation. Evaluate the theory by assigning your rating to each idea.

Concepts under consideration

From chapter 2

2.1   Centrifugal forces of the earth's rotation are responsible for the oblate ellipsoidal shape of the geoid by thrusting the ocean waters toward the equator.

2.2   Any shift in the earth's axis of rotation results in a corresponding shift of the geoid.

2.3   Any change in the earth's spin rate (length-of-day) changes the magnitude of equatorial bulge by shifting ocean waters between the poles and the equator.

2.4   Any weight moved toward the equator (further from the earth's axis) causes our planet to spin slower, while weights moved toward the pole cause it to spin faster.

2.5   If the North Pole shifts one degree toward the equator along any meridian (shift meridian), all points along the shift meridian shift one degree. The changes in geoid elevation for any point along the shift meridian can be determined simply by calculating the difference in latitude radius of the point before and after the axis shift (assuming a rigid crust).

2.6   The two quadrants in the direction of pole movement (referred to in this book as negative quadrants) experience drainage (higher elevations) because of the lower sea levels (reduced geoid radius). The opposite quadrants (referred to as positive quadrants) experience flooding (lower elevations) because of the higher sea levels (increased geoid radius).

2.7   Evidence that thick sedimentary beds were deposited during periods of continental flooding can be explained by axis shifts even without consideration of crustal gymnastics.

2.8   The gyroscopic stabilizing effect of the earth's oblate ellipsoid shape is altered as the oceans and magma sea respond to axis movements. Consequently, the poles could eventually move to today's equator.

2.9   Since early in earth's history, the amount of water has not fluctuated enough to account for continental flooding during geologic periods when thick sedimentary beds were deposited.

From chapter 3

3.1   The same centrifugal forces that act on the ocean waters also act on the earth's interior (the ductal upper mantle and fluid lower mantle, referred to as the magma seas in this book) to establish the oblate ellipsoidal shape of the lithosphere.

3.2 If our planet started as a spinning fiery ball, its shape was also an oblate ellipsoid before the crust froze.

3.3 If our planet formed from cold galactic materials, heat from collision and radioactive decay caused it to melt before re-freezing to form an igneous rock crust.

3.4 Any earth axis movement will rotate the oblate ellipsoid of the magma seas. Confined within the crustal shell, the magma will respond to an axis shift by flowing away from the negative quadrants (lowering the magma pressure acting on the crust) and toward the positive quadrants (increasing the magma pressure acting on the crust).

3.5 The magma seas, confined within the crustal shell, can be considered as a closed hydraulic system in the transfer of fluid pressures. When one area of crust is pushed down into the magma sea another area is elevated by an equivalent amount like in a waterbed. Consequently, after each cycle of axis shift and magma response the earth starts with a near balanced system. Theoretically, a series of axis shifts could overcome the gyroscopic stability of the equatorial bulge and move the poles to the equator.

3.6 Crustal tension in the positive quadrants, caused by the crust having to span the higher magma seas, opens crustal fractures to facilitate magma intrusions and is responsible for classic geologic features, such as dikes, sills, batholiths, laccoliths, mountain ranges, regional uplifts, and volcanic eruptions.

3.7 Crustal compression in the negative quadrants caused by the crust forced into a shorter span, is responsible for classic geologic features, such as anticlines, synclines, overthrust belts, and subduction zones.

3.8 The amount of crustal stretching and shortening produced by an axis shift can be determined by calculating a family of great circle half-circumferences. The greatest change in crustal length resulting from an axis shift occurs on the great circle that intercepts the shift meridian at 45 degrees latitude.

3.9 Any differential in tension or compression within the crust causes shear stresses. All seismic vibrations (earthquakes) are the result of crustal movements when shear stresses are released.

3.10 The strength of earthquakes is a function of the friction lock on the fault surface.

3.11 Crustal rigidity allows the shear stresses to be transferred to areas of crustal weakness where the rupture occurs. The primary shear stress relief regions are the plate edges and transverse fracture zones associated with the midocean ridges.

3.12 Changes in magma pressure resulting from an axis shift raises and lowers regional areas of the crust like an elevator platform relative to the geoid. The Colorado Plateau Region has been raised and lowered many times, leaving the sedimentary beds essentially horizontal.

3.13 Thermal expansion of the earth's magma interior is insufficient to account for the vertical fluctuations found in the earth's crust.

From chapter 4

4.1 The amplitude variation of the earth's axis wobble (diameter of the cyclic circular trace) is a function of the phase agreement between the 14.29-month Chandler, 12-month annual, and other minor cyclic and non-cyclic wobble components.

4.2   The 42 year peaks in the amplitude of the earth's wobble are a product of the phase agreement primarily between the Chandler and annual components.

4.3   The annual wobble component is attributable to seasonal weight shifts on our planet.

4.4   The Chandler wobble component is attributable to cumulative weight shifts. The eccentric loading on today's earth produces a wobble cycle of about 14.29 months.

4.5   Magnetic pole movements distort paleomagnetic records used to plot tectonic plate movements.

4.6   Magma sea horizontal and vertical current patterns are established by the Coriolis effect of the spinning orb and the heat from nuclear reaction deep within the earth's core.

4.7   Hot spots are the asthenosphere's equivalent of tornadoes, consequently they cannot serve as fixed references for determining tectonic plate movements.

4.8   Warm regions of the crust are caused by cyclonic magma currents convecting heat up from the earth's interior.

From chapter 5

5.1   The MOHO represents an isothermal-pressure boundary (not an isostatic boundary) although, isostatic responses of the crust may temporarily alter the MOHO.

5.2   The centrifugal force created by each element of our spinning orb represents a torque acting on the earth's axis of rotation that can affect its dynamic balance.

5.3   The removal of an object from one location and its placement at a second location can be thought of as a force couple acting on the earth's axis of rotation to alter its unrestrained axis of rotation. Weights moved in the north and south direction at 45 degrees latitude provide the greatest torque on the earth's axis.

5.4   Any isostatic response to crust subsidence, because of increased loading (ice sheets or sedimentary deposits), must either be offset by a companion crustal uplift or increased magma pressure restrained by the crust. The ocean floor, being thinner, will flex in response to variations in isostatic produced pressure.

5.5   Isostatic balance keeps the oceans at nearly a constant level as the ocean crust responds to the weight of water exchanged between oceans and ice sheets.

5.6   Whenever weight shifts cause our planet to move to a new axis of rotation, the kinetic energy of the earth's rotation is tapped to accomplish the work of remodeling the crust.

5.7   The annual wobble cycle components of the earth's axis are caused by seasonal eccentric load shifts of our spinning orb.

5.8   The Chandler wobble cycle component of 14.29 months is caused by eccentric loading of our spinning orb that accumulates and is not relieved by axis drifting.

5.9   Axis drift is caused by the planet trying to reestablish its spin balance.

5.10   The majority of south flowing rivers in the Americas team up with the majority of north flowing rivers in Asia to increase the force couple acting on the earth's axis.

5.11   The high and low density cells of the asthenosphere's circulation pattern provide the greatest long term weight shifts acting on our planet.

5.12   An increase in the eccentric loading of the earth will eventually reach a critical point that triggers the axis surges in its effort to reestablish a spin balance. Weigh shifts resulting from an axis surge creates additional eccentric loading and axis adjustments (the equivalent of earthquake after shocks.

5.13   Tectonic plate movements are Nature's way of retailoring the crust to fit the ever changing oblate ellipsoid of the magma seas.

From chapter 6

6.1   Cataclysmic events that depend on the timing of a combination of contributing factors are common throughout Nature. Axis surges set in motion cataclysmic events during past geologic ages that were more destructive than those observed in recorded history.

From chapter 7

7.1   The geoid's cataclysmic changes created the marine terraces.

7.2   An axis surge triggered the rapid decay (in geologic time) of the ice sheet that covered northern North America and western Europe. It caused an abrupt change in the regional climates around the world. The onset or decay of new ice sheets occurs when an axis surge alters the regional weather by changes in latitude, elevation, and land-ocean distribution (changes in ocean and atmospheric circulation patterns.)

7.3   Each break in the sedimentary bedding represents a cataclysmic axis surge that changed the source and mechanics of deposition.

7.4   Mammoths were quickly frozen and preserved for thousands of years when a cataclysmic axis surge abruptly changed the climate.

7.5   Submarine canyons were carved by conventional processes when they were above sea level. Channels of the submarine canyons were preserved by axis surges that prevented them from being buried by delta sediments as axis drift slowly drowned the canyons.

7.6   Fjords, now thousands of feet below sea level, were carved by glaciers at a time when the land was at or above sea level.

7.7   Axis surges that alter land-ocean boundaries and climates are the primary cause of mass extinction events.

7.8   Magnetic pole movements are a function of changes in the magma circulation patterns as the axis drifts. Magnetic reversals are the product of axis surges abruptly altering the magma circulation patterns within our planet.

7.9   Magma sea tides created by the Chandler wobble combine with earth tides caused by the moon and sun's gravitational pull and shifts in atmospheric loading. They continually massage the planet's crust and are responsible for earthquakes and volcanoes around the world.

From chapter 8

8.1   The Milankovitch Cycles contribute to global climate changes, but axis movements provide the primary driving force.

8.2   Regional climates respond to both surface and deep ocean current patterns that are altered by changes in the ocean floor and land-ocean distribution.

8.3   The depths of sedimentary beds on the continents and continental shelves raises serious questions on the preservation of continents through the cycles that form and divide supercontinents.

Back to chapter 1
1.1 An axis shift of 4.23 degrees away from Denver, Colorado will make Denver a port city (assuming a rigid crust).
1.2 An opposite axis shift of 4.23 degrees toward Denver will make Denver a two-mile high city.

Appendix A ACRONYMS

The following is a short list of acronyms encountered in earth science literature and conversation.

APWP — Apparent Pole Wonder Path.

BIH — Bureau International l'Heure (a scientific organization that tracks the path of the earth's axis as defined by the poles).

C & GS — Coast and Geodetic Survey. An agency of the United States government responsible for establishing and maintaining a geodetic reference system. Also was called NGS (for National Geodetic Survey) for a period of time.

DAT — Dynamic Axis Theory (a theory introduced in this book that explains geological events as a function of the earth's axis of rotation).

DTM — Digital Terrain Model (the earth's surface defined by an array of elevations).

GEOSAT — Satellite based radar altimeter used to measure undulations in the ocean surface.

GPS — Global Positioning Satellite (a satellite system designed to establish geodetic positions).

HAB — HAB Theory (an earth science theory presented by Hugh Auchincloss Brown to explain evidence of cataclysmic events in geologic history).

ILS — International Latitude Service (a scientific organization that tracks the path of the earth's axis as defined by the poles).

IPMS — International Polar Motion Service (a scientific organization that tracks the path of the earth's axis as defined by the poles).

K-T — Cretaceous-Tertiary boundary associated with the mass extinction event that killed the dinosaurs.

l.o.d. — length of day.

Moho — Mohovoricic Discontinuity (the surface between the brittle crust and the pliable mantle.

MYBP — million years before present.

NAVD 29 — North American Vertical Datum adopted in 1929.

NAVD 88 — North American Vertical Datum adopted in 1988.

NAD 27 — North American Datum adopted in 1927, based on Clark Spheroid 1866.

NAD 83 — North American Datum adopted in 1983, based on a new spheroid developed by C & GS.

SEASAT — Satellite based radar altimeter used to measure undulations in the ocean surface.

TPW — True Pole Wander.

UTM — Universal Transverse Mercator (a geodetic reference system adopted by the United States Military, that uses six degree bands of [X,Y, Elevation] Cartesian coordinates.

VLBI — Very Long Base Interferometry (a survey technique that utilizes the radio emissions from distant quasars to establish high precision geodetic positions unaffected by tectonic plate movements).

Appendix B   GEODETIC SUPPORT TABLES

The tables in Appendix B are included to illustrate changes in the geoid and earth's crust resulting from earth axis movements, and to provide a foundation for computations of axis movements. Fundamental to the Dynamic Axis Theory is that the geoid (sea level) is a function of latitude. We can therefore use the changes in spheroid radius to determine geoid changes, if we assume the crust is rigid. Geoid undulations present no problem in determining sea level changes, if we assume the undulations are a function of the planet body.

Table B-1 summarizes the Clark 1866 spheroid radius values [R] (measured in meters from the center of the spheroid) for each degree of latitude. The spheroid radius decreases as a function of latitude. It is at maximum at the equator and at minimum at the poles. To illustrate the effect of a 1-degree axis shift, the radius difference [dR] for each latitude, relative to the next higher latitude, is given in both meters and feet. Note that the dR values are smallest at the equator and poles, with the largest radius change occurring at 45 degrees latitude.

The radius [r] (measured in meters from the earth's rotating axis) of each latitude is also listed. As weights move around on the planet, the radius establishes the centrifugal force magnitude of the moving weight. It is zero at the poles and at maximum at the equator.

Table B-2 lists the great circle half-lengths [C/2] (in meters) for each latitude intercept along the shift meridian. The C/2 value is half of the great circle circumference and represents the length across a quadrant. The differences between adjacent great circle half-lengths [dC] are given in both meters and feet. The smallest dC values are found on great circles that intercept the shift meridian at low and high latitudes—indicating small transverse shear forces. The largest dC value occurs on the great circle that intercepts the shift meridian at 45 degrees latitude—indicating large transverse shear forces.

Table B-3 summarizes the pole motion data with [x] and [y] as arc second offsets to the geodetic reference system. Dates are given as a decimal fraction of the year.

Table B-4 facilitates assessment of the geoid elevation changes for points not on the shift meridian. Values are provided as a percentage of shift meridian changes for each degree away from the shift meridian.

To calculate the change in elevation for any point off the shift meridian:

1. Determine latitude of point in question (degrees).
2. Use Table B-1 to obtain [dR] for the same latitude on the shift meridian.
3. Determine the longitude distance (to the nearest degrees) from the shift meridian meridian for the point in question.

Example: select the point 28 degrees from the shift meridian:
    Follow the top scale across to 20.
    Follow column down until you are opposite the left scale of 8.
    The geoid change for this example is 91% of the geoid change on the shift meridian.

Table **B-5** provides the geoid change values for 189 locations resulting from a theoretical 1-degree axis shift that has the North Pole at (89°N, 70°W). This is where the North Pole will be in about 1.25 million years if the present rate of axis drift continues. The computation of these values assumes the crust is rigid and only the change from the geoid rotation comes into play.

Recognizable geographic locations were selected to illustrate the geoid change. For each location, the present latitude [Lat] and longitude [Long] is given in degrees. For each location, column [SM] gives the geoid change (in feet) for a point at the same latitude on the shift meridian. Column [dLong] lists the longitude distance from the shift meridian in degrees. Table B-4 was used to determine the percentage of shift meridian change given in column [%SM]. Column [dG] gives the geoid change in feet at the selected location.

Table B-5 can easily be used to determine the geoid changes resulting from an axis shift in the opposite direction that puts the North Pole at (89°N, 110°E), simply by reversing the sign for each dG.

Calculations can also be made for smaller or larger axis shifts along the same shift meridian simply by multiplying dG by the degrees of axis shift desired. Thus, 5 x [dG] gives the value for a 5-degree axis shift, and 0.1 x [dG] gives the value for a 0.1 degree axis shift.

Table **B-6** is a special computation that reduces the geoid change values for 1-degree axis shifts (from Table B-1) to 0.73 seconds of arc axis shift. The 0.73 was selected as one of the maximum axis shifts (diameter of axis wobble cycle) recorded to date for a 14-month axis wobble cycle. Geoid change values are given in both centimeters and inches.

TABLE B-1
Spheroid Radii as a Function of Latitude

Lat. = Latitude values in degrees.

Radius [R] = Radius of spheroid as measured from the earth's center based on Clark 1866 spheroid (used for NAD 29)

a = 6,378,206.4 meters (semi-major axis)

b = 6,356,583.8 meters (semi-minor axis)

e = 0.082271854 (eccentricity)

[dR] = Difference in [R] for adjacent latitude radii.

[dR] = Abs (Rn+1 - Rn) where "n" = latitude

[r] = Radius of spheroid as measured from the earth's axis.

contd.

TABLE B-1
Spheroid Radii as a Function of Latitude

Lat.(deg)	Radius [R] (meters)	[dR] (m)	[dR] (ft.)	radius [r] (meters)
0	6378206.4	6.6	22	6378206.4
1	6378199.8	19.7	65	6377228.4
2	6378180.1	32.9	108	6374294.7
3	6378147.2	45.9	151	6369406.2
4	6378101.3	58.8	194	6362564.6
5	6378042.4	71.9	236	6353772.0
6	6377970.5	84.8	278	6343031.3
7	6377885.7	97.5	320	6330345.9
8	6377788.2	110.1	361	6315720.0
9	6377678.1	122.7	402	6299158.3
10	6377555.4	135.0	443	6280666.0
11	6377420.4	147.2	483	6260249.2
12	6377273.2	159.3	523	6237914.5
13	6377113.9	171.2	562	6213668.9
14	6376942.9	182.8	600	6187520.4
15	6376760.2	194.1	637	6159477.4
16	6376566.1	205.2	673	6129548.7
17	6376360.9	216.2	709	6097744.3
18	6376144.7	226.8	744	6064074.0
19	6375917.9	237.1	778	6028548.8
20	6375680.8	247.3	811	5991180.2
21	6375433.5	257.0	843	5951979.9
22	6375176.5	266.5	874	5910960.7
23	6374910.0	275.4	904	5868135.6
24	6374634.3	284.5	933	5823518.2
25	6374349.8	293.0	961	5777122.9
26	6374056.8	301.0	988	5728964.3
27	6373755.8	308.8	1013	5679058.0
28	6373447.0	316.2	1037	5627419.7
29	6373130.8	323.2	1060	5574065.8
30	6372807.6	329.8	1082	5519013.3
31	6372477.8	336.0	1102	5462279.6
32	6372141.8	341.7	1121	5403882.7
33	6371800.1	347.2	1139	5343841.2
34	6371452.9	352.1	1155	5282173.9
35	6371100.8	356.6	1170	5218900.3
36	6370744.2	360.7	1183	5154040.3
37	6370383.5	364.3	1195	5087614.5
38	6370019.2	367.6	1206	5019643.6
39	6369651.6	370.3	1215	4950149.0
40	6369281.3	372.4	1222	4879152.6
41	6368908.7	374.2	1228	4806676.4
42	6368534.2	375.9	1233	4732743.2
43	6368158.3	376.8	1236	4657376.2
44	6367781.5	377.3	1238	4580598.7

contd.

Table B-1. contd.. Spheroid Radii as a Function of Latitude

Lat.(deg)	Radius [R] (meters)	[dR] (m)	[dR] (ft.)	radius [r] (meters)
45	6367404.2	377.3	1238	4502434.7
46	6367026.9	376.9	1237	4422908.5
47	6366650.0	376.0	1234	4342044.9
48	6366274.0	374.6	1229	4259868.8
49	6365899.4	372.8	1223	4176405.8
50	6365526.6	370.6	1216	4091681.6
51	6365156.0	367.8	1207	4005722.5
52	6364788.2	364.7	1197	3918554.9
53	6364423.5	361.0	1185	3830205.7
54	6364062.5	356.8	1171	3740702.1
55	6363705.5	352.5	1156	3650071.5
56	6363353.0	347.6	1140	3558341.8
57	6363005.4	342.3	1123	3465541.1
58	6362663.1	336.5	1104	3371697.7
59	6362326.6	330.3	1084	3276840.4
60	6361996.3	323.8	1062	3180998.2
61	6361672.5	316.7	1039	3084200.0
62	6361355.8	309.5	1015	2986475.6
63	6361046.3	301.6	990	2887854.6
64	6360744.7	293.6	963	2788367.0
65	6360451.1	285.1	935	2688042.8
66	6360166.0	276.0	906	2586912.6
67	6359889.6	267.1	876	2485006.8
68	6359622.5	257.7	845	2382356.5
69	6359364.8	247.9	813	2278992.5
70	6359116.9	237.8	780	2174946.1
71	6358879.1	227.4	746	2070248.5
72	6358651.7	216.7	711	1964931.4
73	6358435.0	205.7	675	1859026.5
74	6358229.1	194.6	638	1752565.5
75	6358034.5	183.2	601	1645580.4
76	6357851.3	171.6	563	1538103.4
77	6357679.7	159.7	524	1430166.8
78	6357520.0	147.7	485	1321802.7
79	6357372.3	135.5	445	1213043.8
80	6357236.8	123.1	404	1103922.6
81	6357113.8	110.5	363	994471.7
82	6357003.3	97.9	321	884723.9
83	6356905.4	85.0	279	774711.9
84	6356820.4	72.1	237	664468.7
85	6356748.3	59.2	194	554027.1
86	6356689.1	46.0	151	443420.2
87	6356643.1	33.0	108	332681.0
88	6356610.1	19.7	65	221842.5
89	6356590.4	6.6	22	110937.8
90	6356583.8	0.0		

TABLE B-2
Semi-Circumference Great Circles as a Function of Latitude

Lat. (deg)	C/2 (meters)	dC (m)	dC (ft)	Lat. (deg.)	C/2 (meters)	dC (m)	dC (ft.)
0	20037726	10	33	43	20021949	591	1939
1	20037716	31	102	44	20021358	592	1942
2	20037685	52	171	45	20020766	593	1946
3	20037633	72	236	46	20020173	591	1939
4	20037561	92	302	47	20019582	590	1936
5	20037469	113	371	48	20018992	588	1929
6	20037356	133	436	49	20018404	585	1920
7	20037223	154	505	50	20017819	582	1910
8	20037069	172	564	51	20017237	577	1893
9	20036897	193	633	52	20016660	572	1877
10	20036704	212	696	53	20016088	566	1857
11	20036492	231	758	54	20015522	561	1841
12	20036261	251	824	55	20014961	553	1814
13	20036010	268	879	56	20014408	545	1788
14	20035742	287	942	57	20013863	537	1762
15	20035455	305	1001	58	20013326	528	1732
16	20035150	322	1057	59	20012798	518	1700
17	20034828	340	1116	60	20012280	508	1667
18	20034488	356	1168	61	20011772	497	1631
19	20034132	372	1221	62	20011275	485	1591
20	20033760	389	1276	63	20010790	474	1555
21	20033371	403	1322	64	20010316	460	1509
22	20032968	419	1375	65	20009856	447	1467
23	20032549	433	1421	66	20009409	434	1424
24	20032116	447	1467	67	20008975	419	1375
25	20031669	460	1509	68	20008556	404	1326
26	20031209	472	1549	69	20008152	389	1276
27	20030737	485	1591	70	20007763	373	1224
28	20030252	497	1631	71	20007390	357	1171
29	20029755	507	1664	72	20007033	339	1112
30	20029248	518	1700	73	20006694	323	1060
31	20028730	528	1732	74	20006371	305	1001
32	20028202	536	1759	75	20006066	288	945
33	20027666	545	1788	76	20005778	269	883
34	20027121	553	1815	77	20005509	250	820
35	20026568	560	1837	78	20005259	232	761
35	20026568	560	1837	79	20005027	212	696
36	20026008	566	1857	80	20004815	193	633
37	20025442	572	1877	81	20004622	174	571
38	20024870	577	1893	82	20004448	153	502
39	20024293	581	1906	83	20004295	133	436
40	20023712	585	1919	84	20004162	114	374
41	20023127	588	1929	85	20004048	92	302
42	20022539	590	1936	86	20003956	73	240

contd.

TABLE B-2 contd.
Semi-Circumference Great Circles as a Function of Latitude

Lat. (deg)	C/2 (meters)	dC (m)	dC (ft)
87	20003883	51	167
88	20003832	31	102
89	20003801	10	33
90	20003791		

Lat. =	Latitude of great circle at its intersection with shift meridian
[C/2] =	Semi-circumference
[m] =	meters
[ft.] =	feet
[dC] =	Change in semi-circumference for each great circle relative to the great circle for the next larger latitude. [dC] = Abs [(C/2)n+1 - (C/2)n, where "n" = Latitude in degrees

TABLE B-3
Sample of the International Latitude Service (ILS) Pole Positions Calculated by the International Polar Motion Service (IPMS) in Misuzawa, Japan

Date	x	y	Date	x	y
1899.806	+0.004	+ 0.071	1901.306	+0.093	+0.008
1899.889	-0.042	+0.045	1901.389	+0.091	-0.056
1899.972	-0.021	+0.035	1901.472	+0.136	0.088
1900.056	+0.066	-0.031	1901.556	+0.102	-0.152
1900.139	+0.044	-0.056	1901.639	+0.064	-0.162
1900.222	+0.027	-0.086	1901.722	+0.016	-0.143
1900.306	+0.014	-0.086	1901.806	-0.038	-0.124
1900.389	-0.041	-0.105	1901.889	-0.091	-0.102
1900.472	-0.032	-0.110	1901.972	-0.115	-0.058
1900.556	-0.042	-0.087	1902.056	-0.104	+0.020
1900.639	-0.066	-0.057	1902.222	-0.039	+0.141
1900.722	-0.074	-0.019	1902.306	+0.042	+0.172
1900.806	-0.073	+0.016	1902.389	+0.113	+0.164
1900.889	-0.075	+0.033	1902.472	+0.191	+0.111
1900.972	-0.025	+0.007	1902.556	+0.213	+0.029
1901.056	0.0	+0.041	1902.639	+0.185	-0.040
1901.139	+0.037	+0.018	1902.722	+0.120	-0.040
1901.222	+0.054	+0.039			

Date =	Year and decimal fraction of a year. Pole positions were recorded on about the 20th of each month.
x =	Geographic coordinate of pole position relative to geodetic reference system spheroid, measured in seconds of arc. Positive [x] measured towards Greenwich Meridian.
y =	Geographic coordinate of pole position relative to geodetic reference system spheroid, measured in seconds of arc. Positive [y] measured towards 90 degrees west longitude.

TABLE B-4

Percentage factors for calculating geoid changes away from shift meridian, based on a parabolic approximation*

	00	10	20	30	40	50	60	70	80	90
0	100	99	95	89	80	69	56	40	21	00
1	100	99	95	88	79	68	54	38	19	
2	100	98	94	87	78	67	53	36	17	
3	100	98	93	87	77	65	51	34	15	
4	100	98	93	86	76	64	49	32	13	
5	100	97	92	85	75	63	48	31	11	
6	100	97	92	84	74	61	46	29	09	
7	99	96	90	82	72	58	43	25	04	
8	99	96	91	83	73	60	45	25	04	
9	99	96	90	81	70	57	41	23	02	

* See Fig. B-1

To calculate the change in elevation for any point off the shift meridian:
1. Determine latitude of point in question (degrees).
2. Use Table B-1 to obtain [dR] for same latitude on the shift meridian.
3. Determine longitude distance (to the nearest degrees) from the shift meridian for the point in question. [Example: select the point 28 degrees from the shift meridian]
Follow the top scale across to 20. Follow column down till you are opposite the left scale of 8. The geoid change for this example is 91% of the geoid change on the shift meridian.

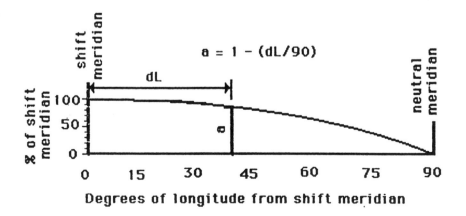

Fig. B-1 Percentage factors for calculating geoid changes based on parabolic approximation [see Table B-4].

TABLE B-5
Change in elevation resulting from a hypothetical one-degree axis shift to (89°N, 70°W)

Location	Lat.	Long. (deg)	[SM] (deg)	[dLong] (feet)	[%S] (deg)	[dG] (feet)
North and Central America						
Pt. Barrow	71N	157W	780	87	07	+55
Nome	64N	166W	990	84	13	-129
Dutch Harbor	53N	167W	1197	83	15	-180
Anchorage	61N	150W	1062	80	21	+223
Juneau	58N	134W	1123	64	49	+719
Prince Rupert	54N	130W	1185	60	56	+664
Seattle	48N	122W	1234	52	67	+827
Portland	45N	123W	1238	53	65	+805
San Francisco	38N	122W	1195	52	67	+621
Los Angeles	34N	118W	1139	48	72	+820
San Diego	32N	117W	1102	47	73	+804
Mazatlan	23N	106W	874	36	84	+734
Guadalajara	21N	105W	811	35	85	+689
San Salvador	13N	89W	523	19	96	+502
Panama	9N	80W	361	10	99	+357
East tipof Honduras	15N	83W	600	13	98	+588
Belize	17N	88W	673	18	96	+646
Merida	21N	90W	811	20	95	+689
Tampico	22N	97W	843	27	91	+767
Brownsville	26N	97W	961	27	91	+825
Corpus Christi	28N	97W	1013	27	91	+922
Houston	30N	95W	1060	25	92	+975
New Orleans	30N	90W	1060	20	95	+1007
Panama City	30N	86W	1060	16	97	+1028
Tampa	28N	82W	1013	12	98	+993
Miami	26N	80W	961	10	99	+951
Jacksonville	30N	82W	1060	12	98	+1039
Charleston	33N	80W	1121	10	99	+1110
Norfolk	37N	76W	1183	6	100	+1183
Washington D.C.	39N	77W	1206	7	99	+1194
New York City	41N	74W	1222	4	100	+1222
Boston	42N	72W	1228	2	100	+1228
Halifax	45N	63W	1238	7	97	+1226
St.John's	47N	53W	1237	17	96	+1188
Hebron	58N	62W	1123	8	99	+1112
Kovik	62N	77W	1039	7	99	+1029
Port Nelson	57N	92W	1140	22	94	+1072
Clyde River	71N	68W	780	2	100	+780
Ft. Ross	72N	94W	746	24	98	+731
MacKenzie Delta	68N	136W	876	66	46	+403
Phoenix	33N	112W	1121	42	78	+874

contd.

Table B-5 contd.

Location	Lat.	Long. (deg)	[SM] (deg)	[dLong] (feet)	[%S] (deg)	[dG] (feet)
Salt Lake City	41N	112W	1222	42	78	+953
Boise	44N	116W	1236	46	74	+915
Casper	43N	106W	1233	36	84	+1036
Denver	40N	105W	1215	35	85	+1033
Albuquerque	35N	107W	1155	37	83	+427
El Paso	32N	106W	1102	36	84	+397
Dallas	33N	97W	1121	27	91	+1020
Oklahoma City	35N	98W	1155	28	90	+1040
Kansas City	39N	95W	1206	25	92	+1110
Omaha	41N	96W	1222	26	92	+1124
Bismark	47N	101W	1237	31	88	+1089
St. Paul	45N	93W	1238	23	93	+1151
Chicago	42N	88W	1228	18	96	+1179
St. Louis	39N	90W	1206	20	95	+1146
Memphis	35N	90W	1155	20	95	+1097
Atlanta	34N	84W	1139	14	98	+1116
Cincinnati	39N	4W	1206	14	98	+1182
Pittsburgh	41N	80W	1222	10	99	+1210
Duluth	47N	97W	1237	22	94	+1163
Sault Ste Marie	46N	84W	1238	14	98	+1213
Detroit	42N	83W	1228	13	98	+1203
Buffalo	43N	79W	1233	9	99	+1221
Quebec	47N	71W	1237	1	100	+1237
Toledo	42N	83W	1228	13	98	+1203
Port Huron	43N	82W	1233	12	98	+1208
St. Lawrence River (70W)	48N	70W	1234	0	100	+1234
Hamilton	43N	80W	1233	10	99	+1221
Kingston	44N	76W	1236	6	100	+1236

Greenland and North Pacific Islands

Location	Lat.	Long. (deg)	[SM] (deg)	[dLong] (feet)	[%S] (deg)	[dG] (feet)
Godthaab	64N	50W	990	20	95	+941
Cape Farewell	60N	44W	1084	26	92	+997
Scoresbysund	71N	23W	780	47	73	+569
Thule	77N	69W	563	1	100	+563
Port-au Prince	18N	72W	709	2	100	+709
San Juan	18N	66W	709	4	100	+709
Kingston	18N	77W	709	7	94	+702
Havana	23N	82W	874	12	98	+857
Bermuda	32N	65W	1102	5	100	+1102
Porta Delgada	38N	26W	1195	44	76	+908
Reykjavik	64N	22W	990	48	72	+713

contd.

Table B-5 contd.

Location	Lat.	Long. (deg)	[SM] (deg)	[dLong] (feet)	[%S] (deg)	[dG] (feet)
Western Europe						
Dublin	53N	6W	1197	64	49	+587
Oslo	60N	11E	1084	81	19	+206
Stockholm	59N	18E	1104	88	04	+44
Copenhagen	56N	12E	1156	82	17	+197
Amsterdam	53N	4E	1197	75	31	+370
Paris	49N	2E	1239	72	36	+442
Madrid	40N	4W	1215	66	46	+559
Lisbon	39N	9W	1206	61	54	+651
Gibralter	36N	6W	1170	64	49	+573
Rome	42N	12E	1228	82	17	+209
Vienna	48N	16E	1234	86	09	+111
Berlin	52N	13E	1207	83	15	+181
Budapest	47N	19E	1237	89	02	+25
Eastern Europe						
Athens	38N	24E	1195	86	09	-108
Istanbul	41N	29E	1222	81	19	-232
Belgrade	45N	21E	1238	89	02	-25
Warsaw	52N	23E	1207	87	07	-84
Helsinki	60N	25E	1084	85	11	-119
Murmansk	69N	33E	845	77	27	-228
Russia						
Moscow	56N	38E	1156	72	36	-416
Rostov	47N	40E	1237	70	40	-495
Baku	40N	50E	1215	60	56	-680
Teshkent	41N	70E	1222	40	80	-978
Chelyabinsk	55N	61E	1171	49	70	-820
Kara	69N	65E	845	45	75	-634
Dikson	73N	80E	711	30	89	-633
Tomsk	56N	85E	1156	25	92	-1064
Irkusk	52N	105E	1207	05	100	-1207
Lena R. Delta	72N	128E	746	18	96	-716
Bodaybo	58N	115E	1123	05	100	-1123
Vadivostok	43N	132E	1233	22	94	-1159
Okha	54N	143E	1185	33	87	-1031
Petropavlovsk	63N	159E	1015	49	70	-711
Uelen	67N	171E	906	61	54	-489
Southern Asia						
Utan Bator	48N	107E	1234	03	100	-1234
Tokyo	36N	140E	1070	30	89	952

contd.

Table B-5 contd.

Location	Lat.	Long. (deg)	[SM] (deg)	[dLong] (feet)	[%S] (deg)	[dG] (feet)
Seoul	38N	127E	1195	17	96	-1075
Peiping	40N	116E	1215	06	100	-1215
Shanghai	31N	22E	1082	12	98	-1060
Canton	23N	113E	874	03	100	-874
Saigon	11N	107E	443	03	100	-443
Singapore	1N	104E	22	06	100	-22
Calcutta	22N	88E	843	22	94	-792
Bombay	19N	73E	744	37	83	-618
Delhi	28N	77E	744	33	87	-881
Kuwait	29N	48E	1037	62	53	-550
Mount Everest	28N	87E	1013	23	93	-942
Aden	13N	45E	523	65	48	-251
North Pacific Islands						
Wake	19N	167E	744	57	60	-446
Midway	28N	177W	1013	67	45	-679
Guam	13N	145E	523	35	85	-445
Honolulu	21N	158W	811	48	72	+584
Okinawa	26N	128E	961	18	96	-923
Manila	15N	121E	600	11	99	-594
Australia, South Pacific and Indian Ocean Islands						
Perth	32S	116E	1102	06	100	+1102
Melbourne	38S	145E	1195	35	85	+1016
Sydney	33S	152E	1121	42	78	+874
Brisbane	28S	153E	1013	43	77	+780
Cape York	11S	142E	443	32	87	+385
Darwin	13S	131E	523	21	95	+497
Jakarta	07S	107E	278	03	100	+278
Papua	10S	148E	402	38	82	+330
Galapagos	0	90W	0	20	95	0
Hobart	43S	147E	1233	37	83	+1023
Christchurch	44S	173E	1236	63	51	+630
Suva	18S	178E	709	68	43	+305
Africa						
Casablanca	33N	8W	1121	62	53	+594
Dakar	15N	17W	600	53	67	+402
Lagos	6N	3E	236	73	36	+85
Luanda	9S	13E	361	83	17	-61
Capetown	34S	18E	709	88	04	-28
Port Elizabeth	34S	26E	1139	84	13	+148
Lourenco Marques	26S	33E	961	77	27	+259
Zanzibar	7S	39E	278	71	38	+106
Mogadiscio	4N	45E	151	65	48	-72
Cairo	30N	31E	1060	79	23	-224

contd.

Table B-5 contd.

Location	Lat.	Long. (deg)	[SM] (deg)	[dLong] (feet)	[%S] (deg)	[dG] (feet)
Tripoli	33N	13E	1121	83	15	+168
Johannesburg	26S	26E	961	84	13	+125
Entebbe	03	2E	0	78	25	0
Lake Chad	13N	14E	523	84	13	+68
South America						
Guayqulil	2S	80W	65	10	99	-64
Lima	12S	77W	483	07	99	-478
Valparaiso	33S	72W	1121	02	100	-1121
Cape Horn	56S	67W	1156	03	100	-1156
Buenos Aires	35S	58W	1155	12	98	-132
Rio deJaneria	23S	43W	874	27	91	-795
Bellem	2S	49W	65	21	95	-62
Georgetown	7N	57W	278	13	98	+272
Caracas	10N	67W	906	03	100	+906
South Atlantic Islands						
St. Helen	16S	6W	637	64	49	-312
Trinidad	20S	29W	728	31	88	-641
S. Sandwich	58S	25W	1123	45	75	-842
S. Shetland	61S	60W	1062	10	99	-1050
Antarctica						
Little America	78S	164W	524	88	04	+24
Robertson Bay	71S	170E	780	60	56	+437
Budd Coast	66S	112E	935	02	100	+935
Seal Bay	73S	12W	711	58	58	-412
Tip Palmer Peninsula	63S	58W	1015	12	98	-995
Mt. Stephenson	70S	70W	813	0	100	-813

[dG] = [SM][%SM]
where:

Location =	point where geoid change is calculated
Lat. =	Latitude of point
Long. =	Longitude of point
[SM] =	Change in geoid if point was on the shift meridian
[dLong] =	Degrees of longitude of point away from shift meridian
[%SM] =	Percent of shift meridian change for point (% taken from Table B-4)
[dG] =	Change in elevation at location = [SM%] x [SM]
(+) =	Lower sea level, higher elevation (drainage)
(-) =	Higher sea level, lower elevation (flooding)

TABLE B-6
Shift meridian geoid changes as a function of latitude for 0.73" axis shift
(Chandler wobble)

Lat. (deg.)	[dR'] (cm)	[dR'] (in)	Lat. (deg.)	[dR'] (cm)	[dR'] (in)
0	0.0	0.00	90	0.0	0.00
5	1.3	0.52	85	1.3	0.52
10	2.6	1.03	80	2.6	1,03
15	3.8	1.51	75	3.8	1.51
20	4.9	1.93	70	4.9	1.93
25	5.9	2.30	65	5.9	2.30
30	6.6	2.61	60	6.6	2.61
35	7.2	2.83	55	7.2	2.83
40	7.5	2.97	50	7.5	2.97
45	7.7	3.01			

Lat. =   Latitude along shift meridian.
[dR'] = Change in Radius for 0.73" axis shift.
          Computed as proportional part of [dR] taken from Table 1
          [dR'] (centimeters) = {[dR](meters) x 0.73 x 100} / [3600]
          [dR'] (inches) =   {[dR] (feet) x 0.73 x 12]} / [3600]

Appendix C   CENTRIFUGAL FORCES AND MOMENTS

Much to the consternation of my mother, I would swing a bucket of milk in a big arc over my head on my way home from the barn. Usually, not a drop of milk spilled thanks to the centrifugal force that held the milk in the bucket. On a couple of occasions, however, the smooth path of the swinging bucket was interrupted: once by bumping the cloth line, and another time by bumping my leg. Of course, the cats were more pleased than my mother.

Equatorial Bulge

The same centrifugal force that kept the milk in the bucket when it was upside down over my head, thrusts the ocean waters toward the equator as the earth spins on its axis. The centrifugal force that acts on each drop of ocean water is perpendicular to the earth's spin axis as shown in (Fig. C-1). The magnitude of centrifugal force [F] is:

$$F = m \times r_\emptyset \times w2$$

where: $m =$ mass under consideration

$r_\emptyset =$ radius of the mass' orbit around the earth's axis of rotation at latitude [Ø]

$w^2 =$ angular velocity of earth's rotation

In the above equation, we may select any unit of mass [m]. If we assume, for now, that the earth's rotation is steady, [w^2] becomes a constant term. This leaves [$r_\emptyset$] as the only variable affecting [F]. Consequently, [F] increases as [$r_\emptyset$] increase from zero at the poles, to its maximum at the equator (Fig. C-2).

For analysis purposes, it is helpful to substitute the normal and tangential force components, [n] and [t], for the centrifugal force [F]. The normal component, [n] acts perpendicular to the earth surface.

$$n = F \times \text{cosine } \emptyset$$

[n] increases from zero at the poles (where both [F] and [cosine Ø] are equal to zero) to its maximum at the equator, where $n = F$. The normal component, [n], is more than offset by the earth's much stronger gravitational attraction, keeping objects from flying away from the earth.

The tangential component, [t], supplies the thrust of ocean waters toward the equator.

$$t = F \times \text{sine } \emptyset$$

[t] is zero at the poles where $F = $ zero, and at the equator where sine $\emptyset = $ zero. It reaches its maximum influence at 45-degrees latitude.

On our spinning planet, the ocean waters are in equilibrium when the tangential gravity component that pulls the 13.5 mile deep equatorial bulge waters toward the poles equals the tangential component [t] of the centrifugal force. In the case of a spinning bucket of milk, the outward centrifugal force was countered by an inward centripetal force provided by my arm, the bucket handle, and the bucket. The milk, relative to the bucket, was in equilibrium.

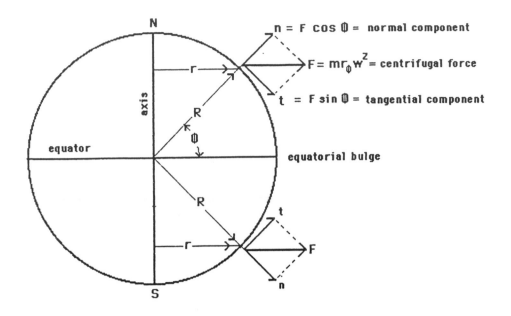

Fig. C-1 Centrifugal forces creating the equatorial bulge.

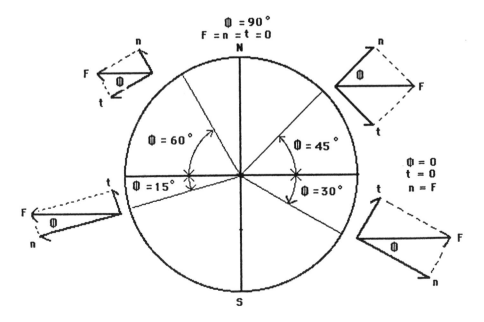

Fig. C-2 Changes in centrifugal forces components from poles to equator.

Any movement of the earth's axis causes a shift in the latitudes [Ø], except along the neutral equatorial and neutral meridional great circles. When the latitude changes, so does the centrifugal force and its thrust component [t]. The magnitude of thrust [t] component is also a function of the earth's rate of rotation [w]. A decrease in earth's spin rate (longer day) decreases the equatorial bulge, while an increased spin rate would increase the bulge.

Moments

The moments of interest are not measured in seconds and minutes, but in pound-feet, ton-miles, or some other weight-distance units. For our purposes the units are not important. In analytical mechanics, the moment [M] is defined as the product of a force and its moment arm.

$$M = F \times d$$

where: [M] = moment (or torque) acting eccentric to equatorial plane and cause a shift of the earth's axis.

[F] = force acting on a body

[d] = moment arm (distance of the acting force from its reference axis. For our spinning planet, the reference axis passes through earth's center and perpendicular to the spin axis). See Fig. C-3.

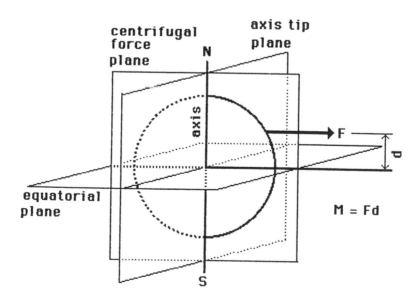

Fig. C-3 Moment arm: perspective view of a force [F] acting eccentric to the equatorial plane.

To illustrate the role of moments, consider that a 50 lb son can balance his 200 lb father on a seesaw, if the boy sits four times as far from the fulcrum as his father (Fig. C-4). In this static illustration, the boy produces a clockwise moment ([MB] = [50] x [4d] = [200d] about the fulcrum) that is balanced by the man's counterclockwise moment ([MM] = [200d]). Forces created by moving weights on our planet can create a torque that acts on the spin axis. The torque amount (moments or a force couple when two units of mass act in opposite directions) is a function of the magnitude of the force and its moment arm.

For our spinning planet the forces to be considered in analyzing the effect of weight shifts are a product of centrifugal forces. For convenience, we will assume that the earth is spinning in perfect dynamic balance. We are interested in the effect any weight shift will have on earth's axis. If we choose the equatorial plane as one of our reference planes, the moment arm [d] is the distance of the weight from the equatorial reference plane as measured parallel to the earth's axis (Fig. C-5).

If we choose for our second reference plane a meridian plane passing through the weight under consideration, the full centrifugal force comes into play. If we include weights (centrifugal forces) not on the selected meridian reference plane, their effective centrifugal force relative to the meridian plane must be reduced by:

$$[FE] = [F] \times [cosine \; \text{ß}].$$

where: [FE] = effective centrifugal force

            [F] =   centrifugal force

            [ß] =   longitude angle between meridian reference plane and the plane
                      of the acting centrifugal force (Fig. C-6).

When we determine the dynamic balance of the spinning earth, the sum of all moments must be considered. We will limit our consideration, for now, to the effect a single weight added or removed from the meridian reference plane.

Other assumptions include:

1. The earth is a sphere of radius [R] (not an ellipsoid).
2. We will ignore any changes in the earth's centroid due to the movement of mass.
3. Paul Bunyan's grandfather uses his big shovel to move Mount Everest to it wherever we need for our illustration.

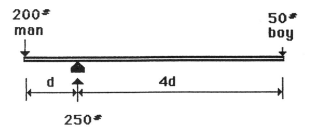

Fig. C-4 Balance moments.

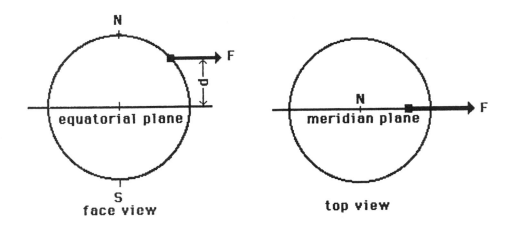

Fig. C-5 Moment arm for force acting on shift meridian and parallel to the equatorial plane.

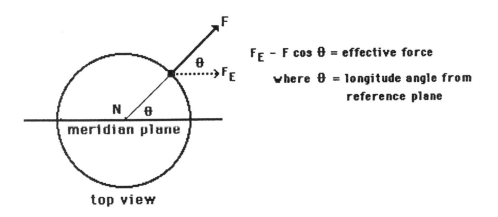

Fig. C-6 Reduced effect of centrifugal forces off the reference plane.

Mount Everest as a Shifting Weight

To illustrate how the movement of weights creates a torque on the earth's rotation axis, consider Mount Everest the moveable mass (Fig. C-7). Applying the equation used earlier:

$$F_E = m_E \times r_{28} \times w^2$$

where:
$[F_E]$ = centrifugal force exerted by Mount Everest

$[m_E]$ = mass (effective weight of Mount Everest)

$[r_{28}]$ = earth's radius at 28 degrees latitude

Using the moment equation: $M_E = F_E \times d$

where: $[M_E]$ = moment exerted on the spin axis by Mount Everest
        [clockwise positive]

   [d] = moment arm for Mount Everest as measured from the
        equatorial reference plane

By our initial assumption, the centrifugal force and resulting moment
illustrated in Fig. C-7, are part of the dynamic force system that is needed to maintain the

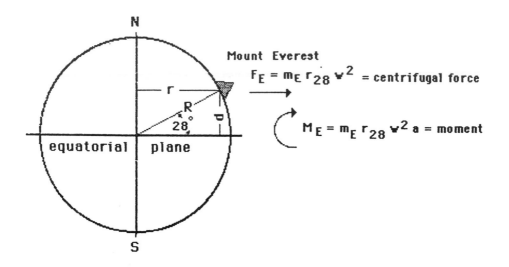

Fig. C-7 Centrifugal force and moment contributed by Mount Everest land mass.

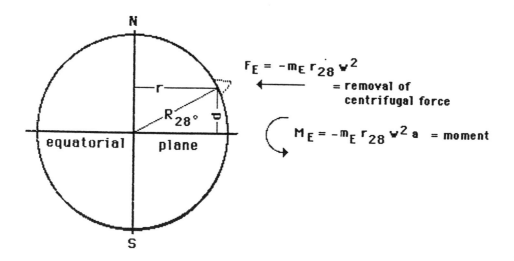

Fig. C-8 Counter force and moment resulting from removal of Mount Everest land mass.

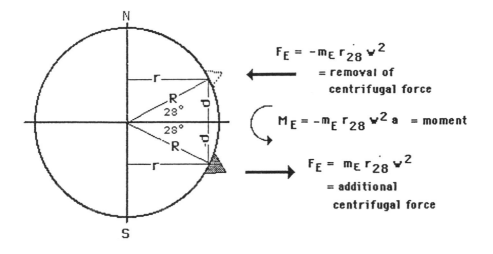

Fig. C-9 Forces resulting from moving Mount Everest to the southern hemisphere.

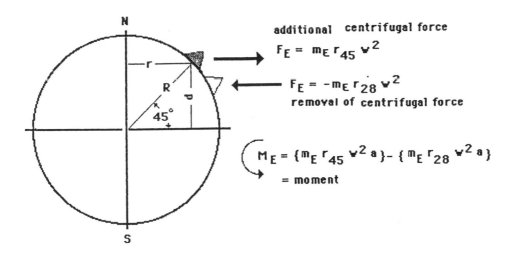

Fig. C-10 Forces resulting from moving Mount Everest to 45° north latitude.

earth's spin balance. By removing Mount Everest, as shown in Fig. C-8, we have created an equal, but opposite centrifugal force and moment (counterclockwise) that upsets the dynamic balance.

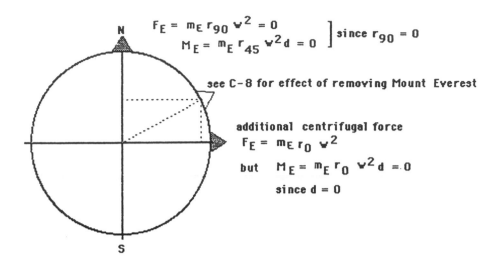

Fig. C-11 Forces resulting from moving Mount Everest either to the pole or to the equator.

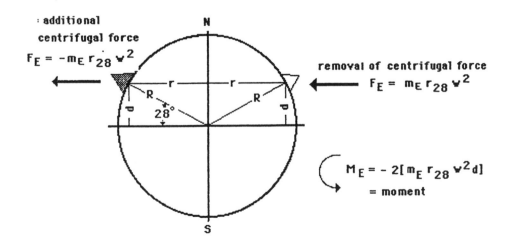

Fig. C-12 Forces resulting from moving Mount Everest 180° on same latitude.

If we now position our hypothetical Mount Everest due south on the 28 degree south latitude, as shown in Fig. C-9, we will double the counterclockwise (negative) moment by doubling the moment arm [d].

By moving Mount Everest due north to 45° north latitude, as shown in Fig. C-10, the resulting moment becomes the algebraic sum of the positive moment for its removal from 28° north latitude, and the negative moment from its placement at 45° north latitude.

Fig. C-10 is typical of weight shifts that occur by erosion (or growth and decay of an ice sheet) in that weights are removed from one location and added to another. (Please note that [d], for weights north and south of the equator, increases as a sine function of the latitude, while [r] decreases as a cosine function.)

Fig. C-11 shows the moment created when Mount Everest is positioned at the poles or on the equator. Since [r] becomes zero at 90° latitude, both the centrifugal force and moment are reduced to zero. At the equator, the centri-fugal force is at its maximum, with [r] = [R]. The moment, however, reduces to zero since [d] = [0]. The resulting torque acting on the spin axis is the same as shown in Fig. C-7 (the only torque is produced by removing Mount Everest.

In our final illustration (Fig. C-12), we have assumed that Mount Everest is moved 180° along 28° north latitude. The resulting moment is positive, consequently, doubling the clockwise moment acting on the axis by Mount Everest today.

Computer Modeling of Weight Shift Moments

The torque applied to the earth's axis is the algebraic sum of all moments resulting from weight shifts. A computer modeling program has to be capable of addressing global weight shifting patterns over time. The first step in establishing a computer program is to divide the earth's surface into a systematic grid of cells. Choosing small cells increases the resolution, but also increases computer capacity requirements and costs.

The next step is to assign values to each cell indicating the gain or loss of weight. For cells subject to more erosion than deposition, a negative value representing the net amount of erosion (such as tons per year, decade, century, etc.). For cells that gain from sediment deposits, positive values are assigned. Judgment is needed to assign the weight shift values for cells that have both erosion and deposition. Similar weight shift factors are needed for changes in glacial ice, lake and reservoir capacities, etc. Short range (semiannual or shorter) computations may involve weight shift values assigned for atmospheric and ocean currents.

One of several approaches to determine the additive torque of all weight shifts is to determine the effective force for each cell as a function of any selected meridian plane. The moment or force couple for each cell is the effective force times its moment arm. The resulting torque acting on the axis is the sum of all cell torques. In this approach, the critical meridian is found by trial and error.

Long term calculations (for past ages) involve weight shifts, such as the North America and Europe's ice sheet growth and decay, the erosion that carved the Colorado River basin and removed the sedimentary cap off of the Rocky Mountains, the flood lava that created the Columbia Plateau and Deccan Traps, and the uplift of the Himalayas. Working with short, medium and long range known and hypothetical weight shifts should give scientists a far better feel for the Dynamic Axis Theory.

Appendix D   AXIS SHIFTING AND EARTHQUAKES

The earth quakes when one crustal block suddenly moves relative to its neighbor. The physical vibrations are recorded on seismographs. The shock wave intensity depends on the strength of the friction lock that is resisting the crustal fracture. Factors that determine the friction lock include the strength of the crust material, the physical roughness of the fault surface, and the compression force holding the blocks together. In other words, the earthquake intensity depends on the crust's ability to store energy for release at some critical point. Albert's experience in the following scenario illustrates the concepts of friction locks.

Albert was asked by his teacher to go to the school library and bring back all the books he could find about analytical mechanics. The teacher suggested he use the empty orange carton on the table at the back of the room. When Albert arrived at the library, he dropped the box on the floor and started his search for books. As he moved along the shelves, he shoved the nearly empty box along the tile floor with his foot. Before long, however, the book weights in the box created such a resistance that Albert was no longer able to move the box with a simple push of his foot. He observed that each book added weight to the box to increase the friction lock between the box and the floor (Fig. D-1). To add to his problems, when Albert bent down to push the box with his hands he shifted some of his weight to the box. This added to the friction lock of the box at the same time it reduced the friction lock between his feet and the floor. Consequently, when he pushed on the box, his feet were as likely to slide as the box. Albert decided to tie a rope around the box so he could pull it (Fig. D-2). It worked, and Albert continued adding books until the box was full. The vertical component of the rope pulling the box offset part of the weight of the box and slightly reduced the friction lock.

Albert's next problem was to get the box back to the classroom in another building. When he pulled the box into the hallway outside the library, the box hit a wet spot that made the box of books much easier to move. In this case, the water served as a lubricant that reduced the friction lock. Albert finally reached the front door and got his box of books out on the concrete sidewalk. He then found it was almost impossible for him to move the box. The increased friction lock was caused by the concrete finish of the sidewalk being much rougher than the tile floor of the library. Fortunately, his friend Bonnie came by and offered to help pull on the rope. With two people pulling on the box, they had little difficulty for the next 50 feet. When they reached the corner of the library building, they could see that a shortcut across some bare ground (which had been a staging area for the construction project) would save them several feet of travel back to the classroom. Again the box of books became near motionless as the two tried to pull it across the bare ground. The friction lock of the rough ground was even greater than the concrete sidewalk. As luck would have it, Carl came by and offered to carry some of the books to see if that would make it any easier for Albert and Bonnie to pull the box. When Carl loaded his arms with books, the friction lock was reduced and the books were soon in the classroom, but not before the bottom fell out of the box just as they reached their destination.

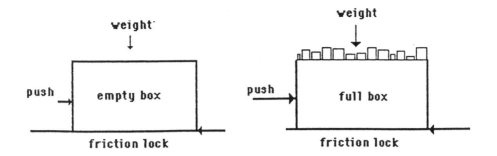

Fig. D-1 Friction lock between box and floor increases as a function of the weight of the box.

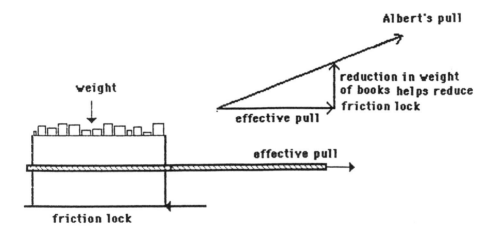

Fig. D-2 Pull equivalent to push force.

In the above scenario, gravity provided the force that created the friction lock. Between the earth's crustal blocks, gravity usually plays a smaller role while horizontal forces that drive tectonic plates are the major players. To determine the friction lock along a fault line between any two crustal blocks, it is necessary to know the direction and magnitude of the forces holding the blocks together. For this, it is also necessary to convert the acting force into its perpendicular (compression) and parallel (shear) component forces acting on the fault surface. It is the perpendicular or normal force component acting on the fault surface that serves the same function as gravity did for Albert's box of library books. The shearing force component parallel to the fault surface must be strong enough to break the friction lock before slippage takes place just as it was for Albert and Bonnie pulling their box with a rope.

Negative Quadrant Compression and Dip Shear

The following analysis of compression forces acting between crustal blocks is an over-simplification intended only to illustrate some fundamental interactions.

Figs. D-3-D-6 show how the horizontal compressive forces acting on the earth's crust can be replaced by normal and shear force components. The magnitude of the normal and shear force components depends on the angle between the fault surface and the compression force. How the crust responds depends in part on the support provided by the magma seas—simple dip slip, plate subduction, or overthrusting plates.

In Fig. D-3, the fault face is perpendicular to the compressive force. The strike normal force [SN] equals the compression force. In this case no shear stress is generated. Nothing happens, unless the compressive forces become so large that the crust crumbles as a rock squeezed in a vise. In the negative quadrants, the negative magma pressure creates a magma void as the magma flows away to the positive quadrant and fails to support for the crust. Before the crust can rest on the lowered magma sea, it is compressed to fit into the shorter span. The notion of a magma void developing is dependent on the crust's ability to span the area of reduced magma pressure. If it exists, it may be just a momentary event before the crust fails. When the crust subsides to become supported by the magma, the negative pressure becomes neutral.

The friction lock along the vertical fault face is similar to the bond created between surfaces when a juggler catches a block of wood between two other blocks, or the keystone of a brick arch.

As the angle of the fault surface deviates from the vertical (decreasing dip angle), a small dip shear force component [ds] is created along with a slightly smaller dip normal force component [dn] (Fig. D-4).

The dip normal force is:
$$[dn] = [SN] [cosine \; Ø]$$
The dip shear force is:
$$[ds] = [SN] [sine \; Ø]$$

Figs. D-5 and D-6 show continued reduction of the dip angle which results in a progressively greater dip shear force component with a decreasing dip normal component.

The weight [w] of the upper block also adds a component force to both the weight normal [wn] and weight shear [ws] forces. If temporary negative magma pressure exists, the right hand block can sink down into the magma sea. When the sinking block consumes the magma void, the magma pressure becomes neutral as indicated in Figs. D-4-D-7 (just enough to support the crust). If the right block is pushed deeper into the magma sea like a plunger in a hydraulic device, the magma pressure becomes positive (possibly enough pressure to push magma up into crustal fractures in the crust or bow up adjacent areas of crust, or even trigger a volcanic eruption).

The normal force responsible for the friction lock along the fault is the sum of [dn] and [wn]. Figs. D-3-D-6 reveal that [dn] decreases as [ds] increases, but [wn] increases as [ws] decreases. It is the friction lock, provided primarily by the dip normal force, that determines how much dip shear force is necessary to create a vertical slippage (earthquake) along the fault.

The greater the magma void, the greater the chance that the right block of Fig. D-6 will curve down and become a subducting plate. If the magma can support the crust, the left block will be pushed up over the right block as an overthrust belt. Elevating the left block allows subsidence of the right block to maintaining magma sea pressure equilibrium.

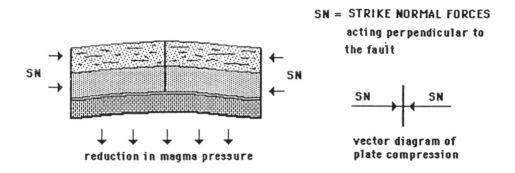

SN = STRIKE NORMAL FORCES
acting perpendicular to
the fault

vector diagram of
plate compression

Fig. D-3. Compressive crustal forces acting on vertical crustal fault.

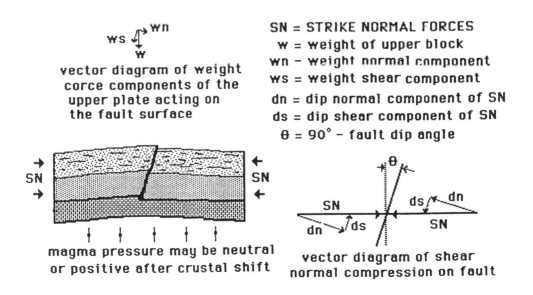

SN = STRIKE NORMAL FORCES
w = weight of upper block
wn – weight normal component
ws = weight shear component
dn = dip normal component of SN
ds = dip shear component of SN
θ = 90° – fault dip angle

vector diagram of weight
corce components of the
upper plate acting on
the fault surface

magma pressure may be neutral
or positive after crustal shift

vector diagram of shear
normal compression on fault

Fig. D-4. Compressive crustal forces acting on near vertical crustal fault.

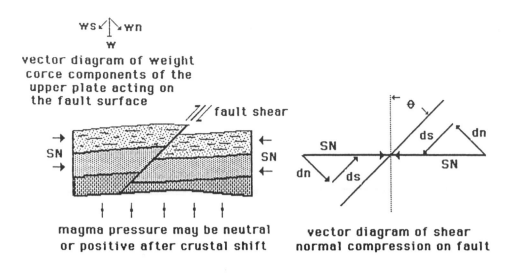

Fig. D-5 Compressive crustal forces acting on a diagonal fault.

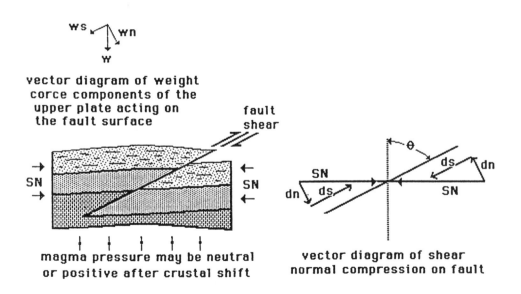

Fig. D-6 Compressive crustal forces acting on an acute diagonal fault and vector diagram. Magma pressure determines if rupture creates a subduction zone or an overthrust plate.

Although a subducting plate can dive down at a sharp angle, the weight of an overthrust plate will not allow it continue skyward.

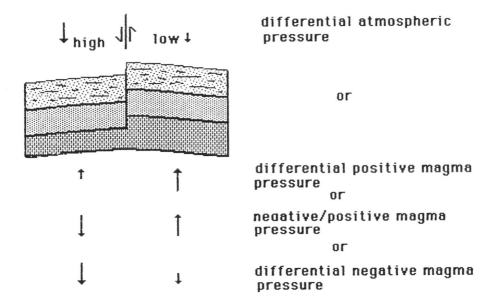

differential atmospheric
pressure

or

differential positive magma
pressure
                    or
negative/positive magma
pressure
                    or
differential negative magma
pressure

Fig. D-7 Vertical shear in crust created by differential atmospheric or magma pressure.

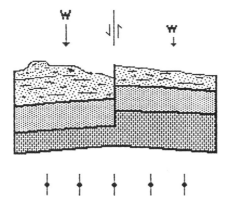

Fig. D-8 Vertical shear on fault resulting from differential crustal loading.

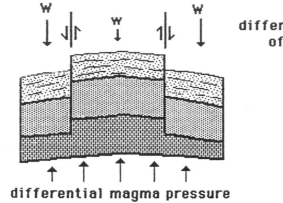

Fig. D-9 Block mountain dual rupture.

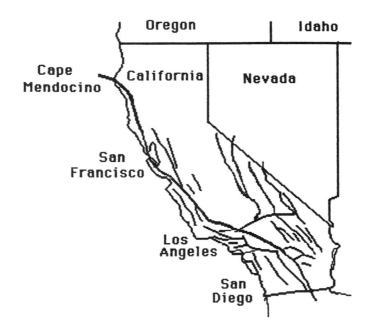

Fig. D-10. Schematic of major faults in California.

Tension and the Positive Quadrant

An axis shift that moves the spin axis away from the fault, as illustrated in Figs. D-5-D-7 (introducing positive quadrant forces) will reduce the plate movement and its component vectors. An axis shift sufficiently large to place the crust under tension would reverse the vector arrows shown except for the weight force. It would be an oversimplification to say that the block would simply pull apart along an existing fracture. An earthquake in this general scenario will be of very low magnitude, since the friction lock is limited to the weight of the upper block being supported by the lower block. The magma under increased pressure of the positive quadrant would push up to fill any openings created by the crustal tension.

Since Nature often likes complexity, we will probably never find a fault surface as illustrated above. It is very doubtful that a single fracture surface ever extends completely through the crust. Large dip slip faults tend to be slightly curved with the upper part being steeper and the lower part at a more acute angle. Consequently, the vector components vary across the fault surface.

Also, very large faults like the San Andreas will rupture in segments rather than all at once. Think of a stiff wire brush being pushed slowly across a rough surface. As stress builds in each wire of the brush, the wire yields by bending slightly. Each wire is released on its time schedule as the accumulated stress reaches a breaking point. Although segments of the earth's crust are not as independent as the individual wires of the brush, a similar stress and release sequence occurs over the fault surface.

Vertical Forces Exerted by Magma Seas

In the above paragraphs, we were looking at the horizontal forces created by compression in the negative quadrant or tension in the positive quadrant. Compression is a product of the reduced crustal span of the negative quadrant. Tension is a product of increased magma sea pressure (higher magma seas), that stretches the crustal skin. We also need to examine differential vertical forces acting directly on the crust by variations in atmosphere loading (high and low pressure cells) and magma pressure. Any local variations in the vertical vectors acting on the crust will create a vertical shear. Fig. D-7 presents four differential pressure situations, including one showing loading by the atmosphere and three indicating variations of magma sea pressure on opposite sides of a fault.

In Fig. D-8, the shear force along the fault surface results from a difference in the weights of the abutting blocks. This may exist between the thick continental crust and the thinner oceanic crust when sediments are deposited on the continental shelves. Weight differentials are also the controlling factor in Fig. D-9, where the center block segment has broken free from the crustal blocks on both sides. In this case, a uniform magma pressure lifts the smaller crustal block that provides the least resistance, resulting in a block mountain.

Strike Shear Forces

In Figs. D-3 -D6, it was assumed that the plane of the compression force was acting at right angles to the strike of the fault surface. Since most crustal ruptures occur along existing faults (relative weaknesses), it is a rare that the compression force provided by a moving tectonic plate acts at right angle to the fault strike. From maps showing a plan view of faults (strike alignments) in a local or regional area, it is easy to see that a single driving force (vector of tectonic plate movement) will intercept several faults and segments of the same fault at different angles (Fig. D-10). The family of faults associated with a fault zone often have a general direction agreement but are far from parallel.

Using the plan view of a fault, we can use the same type of vector analysis employed in Figs. D-3-D-6, this time using Fig. D-11 (a to c). Assume for the moment that tectonic plate motions [PM] represent the horizontal driving force. The normal and shear components of the driving force depend on the angle a vertical fault makes with the plate movement. When the fault surface is not vertical, the forces must be further reduced into components dn and ds, as discussed above using Figs. D-3-D-6.

The shear force acting on the fault surface is the sum of the dip shear and strike shear = [ds2 + SS2]1/2, plus any shear forces induced by vertical loading on the crust.

Forces Along the San Andreas Fault Zone

Strike Shear Faults

The San Andreas Fault marks the boundary between the north-northwestern moving Pacific Plate, and the west-northwest moving North American Plate. The shearing action between the two plates is responsible for many of the earthquakes plaguing residents of California. For illustration purposes, assume that the San Andreas Fault is a vertical

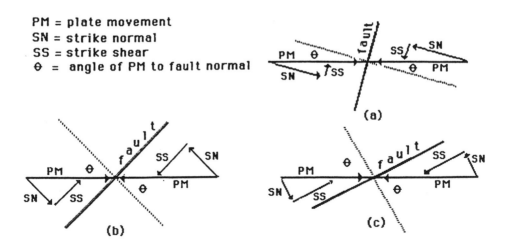

Fig. D-11 Normal and shear components of plate motion are a function of the angle between the line of plate movement and the perpendicular to the fault.

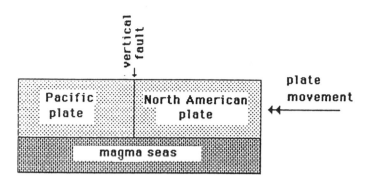

Fig. D-12 Schematic cross section of San Andreas Fault.

surface as shown in Fig. D-12. Also assume that the driving force behind the Pacific Plate movement is parallel to the San Andreas Fault, and the driving force for the North American plate is due west. Since the driving force acting on the Pacific plate, in this illustration, has no strike normal component pushing it towards the North American Plate, it does not contribute to the friction lock that must be overcome before rupture.

The strike normal component provided by the North American Plate, which is directed toward the Pacific Plate, is responsible for the friction lock, as shown in Fig. D-13. If we had assumed the Pacific Plate movement was not parallel to the San Andreas Fault in our illustration, the effective strike normal force would be the algebraic sum of the contributions of the two plates. The relatively small strike shear component of force provided by the North American Plate acts in the same direction as the Pacific Plate strike shear component. The resulting strike shear force between the plates is the algebraic difference of the two strike shear components. In this illustration the contribution by the North American Plate reduces the total strike shear between the plates. When the strike shear stresses along any segment of the San Andreas Fault resulting from the relative movement of the plates exceeds the friction lock, the fault ruptures.

Think of the North American Plate as being a big defensive lineman trying to get to the quarterback, China. For the North American Plate to reach the quarterback, it must go over, around, or through the even bigger Pacific Plate offensive lineman. The role of the Pacific Plate on this particular play is not just to stay between the North American Plate and China, but to be a pulling guard to the left. It is difficult to assign absolute force vectors, but for this scenario we are assuming that the North America Plate is attempting to drive due west. With arms of the opposing linemen partially locked, any lateral movement of the Pacific Plate to the north will tend to drag the North American Plate along with it. Any westward push by the North American Plate would try to deflect the Pacific Plate's path slightly west of its assumed north-northwest travel. In plate tectonics, as in football, the physical movements between offensive and defensive lineman in collision do not tell the whole story about the magnitude and direction of applied forces. In fact, the movement direction may be 180 degrees opposite the force being applied. Ask the lineman sitting on the turf trying to decide, if it was a Mack truck or a bulldozer that hit him.

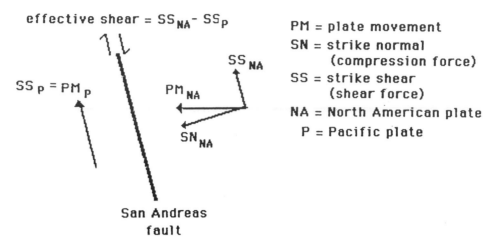

effective shear = $SS_{NA}- SS_P$

$SS_P = PM_P$ PM_{NA} SS_{NA}

SN_{NA}

San Andreas
fault

PM = plate movement
SN = strike normal
        (compression force)
SS = strike shear
        (shear force)
NA = North American plate
 P = Pacific plate

Fig. D-13. San Andreas fault force diagram.

Fig. D-14. San Andreas Fault alignment.

Our assumption that the Pacific Plate driving force and movement is parallel to the San Andreas Fault (Fig. D-13) is obviously erroneous, when we look at a map showing the fault (Fig. D-14). The fault line enters the United States from the Gulf of California and passes to the east of Los Angeles before swinging slightly to the west past the Palmdale region. Another slight turn directs the fault alignment north past the west side of San Francisco. A sharper turn to the west occurs as the fault leaves the mainland near Cape Mendocino. If we assume that the plates move as a unit, it is obvious that the strike shear and strike normal force components will be different for each fault segment within the San Andreas fault zone. Since the strike shear and strike normal forces will vary at every turn in the San Andreas Fault, the friction lock also varies between fault segments. The difference in friction lock is evident between the section near Parkfield, California where the San Andreas Fault moves in more frequent and less intense surges than the section near San Francisco. Remember also that the San Andreas fault zone is composed of several faults, not a single fault.

Deviations in the alignment of the San Andreas Fault from a straight line can also be thought of as the roughness between the two surfaces. This is an obstacle that must be overcome before the plates can move relative to each other. When spurs and knobs of crust jet out from one plate into the path of the other, they are subject to being sheared off and becoming foreign terrains. The freed block of crust is eventually fused onto one of the crustal plate at some other place. Much of the western North American is made up of foreign terrains that have, either been peeled off the side of the North moving Pacific plate, or islands sheared off the top of a subducting plate.

Dip Shear Faults

Fig. D-13 illustrates that the North American plate's major force component strikes normal to the San Andreas Fault. Since our assumption is based on actual plate movement, it is more than just a force being resisted by the Pacific Plate. Any movement of the North American Plate normal to the fault must be absorbed by some combination of plate subduction, plate overthrust, and crustal folding. The San Andreas Fault, as illustrated in Fig. D-12, with a 90 degree dip (vertical fault force) will not allow westward movement of the North American Plate. Below the surface scar of the San Andreas and companion faults are many non-vertical fault surfaces that are known as dip shear faults, similar to the fault diagrams in Figs. D-4-D-6. In absorbing the westward movement of the North American Plate, vertical displacement occurs. Most earthquakes in the Los Angeles area, such as the Whittier Narrows, Northridge and Mexico City earthquake, were subsurface dip shear earthquakes. The Sierra Nevada and Sierra Madre Mountains are the product of dip strike earthquakes as the North American Plate takes the upper path over the Pacific Plate. When the focus of a dip shear fault is deep, it can still provide surface destruction from the vibration without leaving a linear surface scar. The destructiveness of a dip shear rupture depends on the composition of the crustal materials, as for strike shear ruptures. The magnified destruction associated with the Mexico City and Armenia earthquakes would have been considerably less, had the crust under the city been igneous basement rock.

Transverse Faults Along the Midocean Ridge

Some of the most striking features of the National Geographic physical globe with the ocean floor features are the transverse faults aligned perpendicular to the midocean ridges. The family of parallel transverse faults are nearly straight and extend for thousands of miles out from both sides of the ridge. The transverse faults also feature many major offsets in the midocean ridge and rift valley. As discussed in chapter 3, the transverse faults represent a tailoring alteration to the earth's crust to fit the ever changing magma sea's oblate ellipsoid.

At first glance, one would expect the midocean ridge and the transverse fault zones to be seismically very active. On a world map showing the 6.0+ magnitude earthquakes, however, both the midocean ridges and transverse fault zones are relatively quiet zones. Seismic maps show numerous large earthquakes around the Pacific's Ring of Fire and a line of earthquake activity from the Alps to the Himalayas that are a product of compressive forces. The reason for the low seismic activity is found in the above discussion about the role of friction locks in earthquakes. The midocean ridge involves crustal tension and plate spreading. The crust under tension reduces the friction lock to zero. For the long transverse faults associated with the midocean ridge, the differential shear stress is acting parallel to the driving force on the plate. Consequently, they do not have a strike normal force component to establish a strong friction lock. Being straight, they rupture with less shear stress buildup than for the crooked San Andreas fault.

Initial Earthquake Ruptures

The above discussion has been limited to earthquakes along existing faults. Each fault surface had an original rupture. The original rupture of any fault, of course, represents the stronger friction lock. The original rupture will probably be its strongest This resembles the first crack in a windshield under stress that has the loudest cracking sound for the windshield. Next in magnitude is the extension of an original rupture, as the stresses continue to build. After the initial rupture, it is necessary to rebuild the friction locks before the next large magnitude earthquake.

Laboratory Testing

In strength-of-materials laboratories, engineers test all forms of building materials for their ability to resist compression, tension, and shear forces. Concrete is excellent for its compressive strength, but must depend on steel reinforcements wherever tension forces are encountered. Samples from each batch of concrete mix for a critical structure, e.g., a dam or bridge, are tested to insure that the design strength is being maintained. When Nature assembled the lithosphere, no such quality control measures were employed. Although tests can be made on both tension and compression strengths of individual rocks from the crust, the heterogeneous nature of the crust makes it impossible to establish a dependable constant for predicting earthquakes.

Appendix E   OCEAN AND EARTH TIDES

Few notice it, but it takes less energy to move west than east at sundown, and more energy to move west than east at sunrise because of the sun's gravitational pull. Similarly, less energy is needed to move to the east as the full moon rises, and to the west as the new moon sets.

In the main text we have discussed some of natural processes that result from a shift of weights on our planet. Some weights are shifted like clockwork. Ocean tides respond to gravitational attraction of the moon as the earth spins on its axis. Two high and two low lunar tides circle the globe about every 25 hours and 50 minutes. The direct tide (tide on the same side of the planet as the moon) is about 5 percent greater than the second high tide (antipodal) that follows 180 degrees in longitude (or 12 hours) later. Of course, the earth's gravitational pull more than offset the gravitational pull of the moon, sun, or other planets as they to pass overhead. For points on earth at 90 degrees to the solar bodies, their gravitational attraction is perpendicular to earth's gravity, thus introducing a tangential force that creates the ocean tides.

In the open seas, the tidal bulge is only one or two feet high. The higher tides, as measured along the continental coasts, are augmented by the ocean waters being thrust against the land. In the Bay of Funday, the local tide is magnified about 10 times to create a tide range over 20 feet. The size and shape of the adjacent ocean body, offshore depths, and the shape of inlets determine the magnitude of the tide range at each gauging station.

The removal of just one foot of ocean water from the Central Pacific (low tide) would fill an inverted cone with a base size equivalent to the state of Rhode Island and a height over 40,000 feet. Realizing that Nature keeps in motion masses even greater than Mount Everest, puts a different perspective on the hypothetical shift of Mount Everest used in Appendix D's illustration of centrifugal forces and force couples.

Force Diagrams for Moon's Gravitational Pull on the Earth's Oceans

Bodies such as the earth and the moon each create a gravitational attraction as a function of their mass, and inversely as a function of the distance between the two bodies. It is the earth's gravitational pull on the moon that holds the moon in orbit around the earth. In turn, the gravitational attraction exerted by the moon tugs at each element of the earth. Although the moon can send fluid tides sailing across the oceans, it cannot move a grain of sand on the ocean beach. The direct crustal response is so minute that direct physical measurements tax today's technology. Oceans, however, respond to create lunar tides as shown in Fig. E-1. In this illustration we are looking at our earth from a perspective above the North Pole. Eight positions, labeled A-H, are selected on the equator to show vector directions of gravitational pull—the gravitational pull [M] is directed toward the moon. Included in Fig. E-1 are vector diagrams for each point showing their normal [n] and tangential [t] vector components. The normal component (perpendicular to the earth's surface), at point C, is more than offset by the earth's own gravitational pull. At point G the two gravitational pulls are additive. Weight conscious people who want to minimize their weight records should only weigh themselves at high noon on a super sensitive scale during the new moon period.

The tangential component creates the tides. The magnitude and direction of the t vector vary as a function of the longitude distance from the moon's nadir (a point on the earth directly below the moon). The moon orbits the earth in the same direction as the earth's spin (but at a slower angular velocity), making the effective cycle period 25 hours 50 minutes, the time it takes the earth to rotate one revolution relative to the moon.

The apex of the moon's gravitational pull occurs slightly behind the moon's nadir because the ocean water viscosity creates a slight lag period. Ocean waters do not rush from A' and E to C to create the bulge. That would be as inefficient as for the last arrival at the theater having to continually excuse themselves as they make their way from the aisle to the innermost open seat. Taking a clue from Nature, the theater patron can accomplish the same results by having each person in the row simply move over one seat, thus opening up a seat on the aisle for the late arrival. Since each element of ocean water needs to travel only a minute distance, the response time for the exchange of positions is dramatically reduced. Ocean tides are created by the rotation of the earth around the moon. The resulting change in sea level (high and low tide) for the equatorial cross-section is shown in Fig. E-2 as viewed from above the North Pole. The direct and antipodal high tides represent water excess at the expense of the low tides' deficit.

A similar diagram can be drawn for the moon's gravitational forces as a function of latitude as shown in Fig. E-3, as we look at earth from a perspective above the equator. The resulting change in sea level for a meridional cross-section is shown in Fig. E-4. Since the moon is positioned always near the equatorial plane, the gravitational pull creates a permanent deficit at the poles. Ocean waters pulled toward the equator by the moon's gravitation are algebraically added to the equatorial bulge created by centrifugal forces of the axis spin.

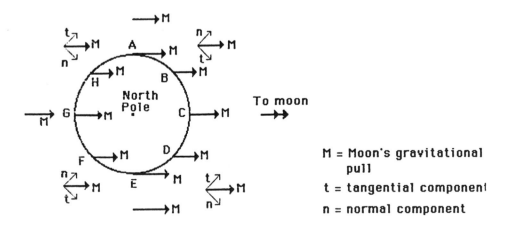

Fig. E-1 Moon's gravitational forces with tangential component [t] and normal component [n] as a function of longitude relative to the moon's nadir.

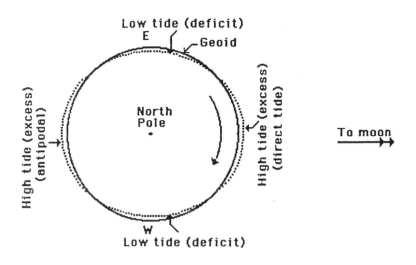

Fig. E-2. Equatorial cross-section of moon's gravitational pull creating high and low tides.

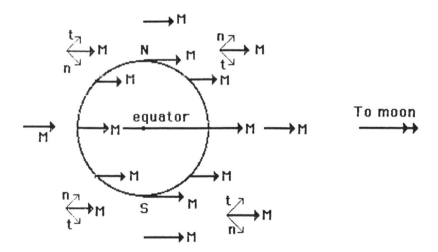

Fig. E-3. Moon's gravitational forces as a function of latitude.

Fig. E-5 illustrates the oscillation created by moon tides along earth meridians. Low tide follows high tide by 90 degrees of longitude and 12 hours 55 minutes later. The budget excess (water) for high tide comes from low tide regions. The polar deficit and equatorial excess noted in Fig. E-4 is not included in Fig. E-5 since the deficit remains nearly constant. As the moon's orbit has a small tilt to the equatorial plane, the apex of the

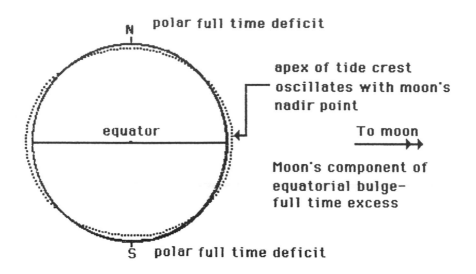

Fig. E-4. Meridional cross-section showing moon's gravitational pull toward the equator.

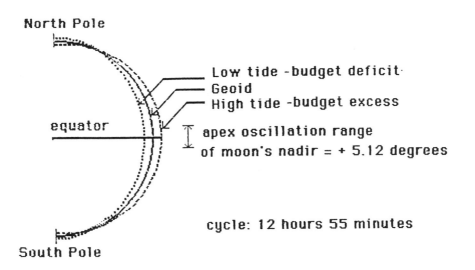

Fig. E-5. Moon tide variations along meridian.

tide crest (along the meridian passing through the moon's nadir) oscillates slightly north and south of the equator following the moon's nadir. Computing the magnitude of geoid changes is beyond the scope of this book. Awareness of their existence is sufficient at this time.

Force Diagrams for Sun's Gravitational Pull on Earth' Oceans

We can use the same diagrams (Figs. E-1 to E-4) to show the effect of the sun's gravitational pull simply by substituting the sun for the moon. The tide magnitude created by the sun is considerably less than that by the moon because the distance between the earth and sun more than offsets the extra mass of the sun. The sun's maximum gravitational attraction presently occurs at perihelion on January 2 when the earth's elliptical orbit passes closest to the sun. The minimum gravitational attraction occurs at aphelion on June 2 when the earth's elliptical orbit is furthest from the sun. Fig. E-6 illustrates the oscillation created by the sun's contribution to ocean tides.

Oscillation of Geoid Due to Chandler Wobble

In chapter 2, we discussed changes in the geoid (maximum along a shift meridian) caused by an axis movement (assuming a rigid crust). When the axis wobbles, the poles trace an irregular circle around the mean pole position. The shift meridian, passing through the instantaneous pole, circles the planet during the wobble cycle. Fig. E-7 illustrates the sea level change from pole to pole along a shift meridian as it circles the globe. The solid line represents the mean geoid. As the North Pole crosses the 90 degree west longitude (establishing the 90° W and 90° E meridians as shift meridians), the north half of the and the south half of the Eastern Hemisphere become negative quadrants. The opposite quadrants become positive quadrants. After the axis wobble (shift meridian) has moved through 180 degrees of its circular path (about 7.145 months later), the negative and positive quadrants are interchanged.

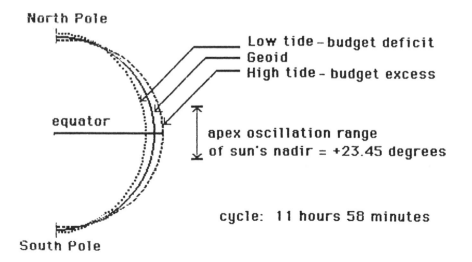

North Pole

Low tide – budget deficit
Geoid
High tide – budget excess

equator

apex oscillation range
of sun's nadir = +23.45 degrees

cycle: 11 hours 58 minutes

South Pole

Fig. E-6. Sun tide variation along meridian.

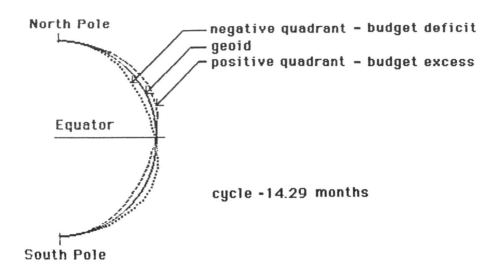

Fig. E-7 Chandler wobble tide.

Change in Geoid Due to Change in the Rate of the Earth's Rotation

The same type of diagram can be produced to show changes in the geoid resulting from a change in the rate of the earth's rotation (change in l.o.d.), see Fig. E-8. The dashed line shows an increased equatorial bulge resulting from an increase in the rate of spin (decrease in l.o.d.). The excess waters that accumulate at the equator are borrowed from the polar region. The dotted line shows a decrease in the equatorial bulge resulting from a decrease in the rate of rotation (increase in the l.o.d.). Cycles in earth's spin rate have been identified at intervals of 2 weeks, 1 month, 2 months, 6 months, annual, biannual, 5 years, and a decade. Any permanent change in rotation rate (lasting for many years), can be treated as a constant for all meridians.

Some estimates of l.o.d. changes over geologic time have pegged the solar day as short as 5 hours. I will leave the computation of the resulting equatorial bulge to others, but the faster spinning orb would indicate our planet was once disc shaped.

Geoid Changes Due to Axis Drift and Surge

For axis drifts and axis surges, assume day one begins coincident with the geodetic reference system axis and pole. The direction of axis drift determines the shift meridian. The path of the North Pole will continue to move in one general direction, if we assume that the weight shifts responsible for an axis shift or surge result from additive erosion patterns, ice sheet growth or decay, or movement of high and low density cells in the magma sea

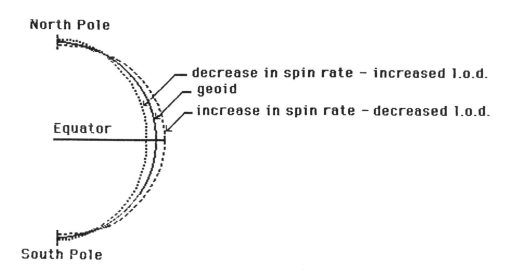

Fig. E-8. Spin rate tides.

current. Axis drift and axis surge establish new origins for considering the axis wobble. The redefined geoid resulting from an axis drift or surge becomes the new reference plane. The negative and positive quadrants continue to circle our planet as a product of the axis wobble.

Combining the Geoid Changes

In the above discussion, we have identified geoid change resulting from the moon's gravitational pull on the oceans, the sun's gravitational pull on the oceans, the earth's rate of rotation, the axis wobble, the axis drift, and the axis surges. Each type of geoid change has internal variations as a function of time, longitude, latitude.

The shape and orientation of the geoid at any time are a product of the algebraic sum of all contributing geoid changes. Combining geoid changes resulting from axis shifts (maximum at 45 degrees latitude) with geoid changes resulting from gravitational attraction (maximum at equator) moves the maximum change to a point from 0-45 degrees latitude (Fig. E-9). The diagram's shaded area is a greatly exaggerated symbolization of the geoid oscillation amplitude. For the magma seas, this explains why the maximum fault lengths in the transverse fracture zones are closer to 30 degrees latitude.

To determine the geoid change for any point on earth at any time, it is necessary to establish the phase influence for each contributing factor, then calculate the algebraic sum. When dealing with the period of recorded history, cataclysmic axis surges discussed in chapter 7 are not a factor.

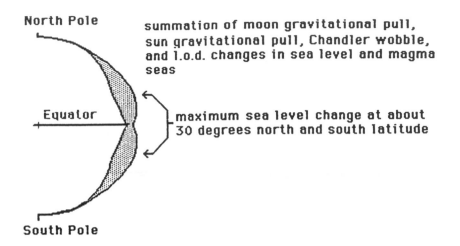

North Pole

summation of moon gravitational pull,
sun gravitational pull, Chandler wobble,
and l.o.d. changes in sea level and magma
seas

Equator

maximum sea level change at about
30 degrees north and south latitude

South Pole

Fig. E-9 Schematic of combined geoid changes.

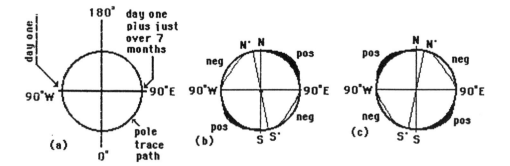

Fig. E-10 Geoid response to axis wobble about GRS.
(a) North Pole trace path concentric about GRS pole
(b) Positive and negative quadrants for day one
(c) Positive and negative quadrants for day one plus just over 7 months

Computing the geoid change amplitude for any location and time is beyond the scope of this book. Even if we tried, it would be suspect since we have not yet considered the effect of magma sea and crustal changes.

Any analysis of the effect of the axis wobble depends on the point of reference. Let us consider three scenarios as we analyze the effects of the Chandler wobble and axis drift on the geoid.

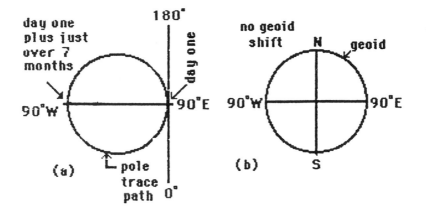

Fig. E-11. Geoid response to axis drift.

First, assume the axis wobble is concentric about the geographic reference system axis. On day one, the North Pole is on 90 degrees west meridian, south of the geodetic reference system pole, and at a distance equal to the radius of the wobble's circular path (Fig. E-10 a). Relative to the GRS pole, the Northwest and Southeast hemispheres are negative quadrants, while the Northeast and Southwest quadrants are positive quadrants (Fig. E-10 b). Just over 7 months later the North Pole is on 90 degrees east meridian, still south of the geodetic reference system pole at a distance equal to the radius of the circular path. The positive and negative quadrants are interchanged (Fig. E-10 c).

Second, assume the axis wobble is concentric about a point on 90 degrees west longitude and at a distance equal to the radius of the pole's circle trace south of the geodetic reference system pole. On day one, the instantaneous North Pole and geodetic reference system North Pole are coincident (Fig. E-11a). All quadrants are neutral (Fig. E-11b). Just over 7 months later, the instantaneous North Pole is on 90 degrees west longitude, and at a distance equal to the diameter of the circular pole trace south of the geodetic reference system pole. As in Fig. E-10 b, the northwest and southeast quadrants are negative while the northeast and southwest quadrants are positive. The magnitude of the geoid change, however, is twice as much as it was for Fig. E-10 b since the total effective axis shift is the diameter of the pole's circular trace.

Third, assume the axis drift has moved the circular pole trace concentric with a point on the 90 degrees west longitude at a distance equal to the diameter of the circular pole trace (not shown). On day one, the instantaneous pole is on the 90 degrees west longitude, at the same point as on day one plus just over 7 months as in our second scenario. Just over 7 months later the North Pole is 2 diameters south on 90 degrees west longitude. The magnitude of the effective geoid change, for this reference, is doubled again.

As we can see, the determination of positive or negative magma sea level resulting from the axis wobble is dependent on what time is chosen as reference point. In Nature there is no natural starting point, consequently, the ideas of positive and negative quadrants are limited to an academic treatment of each axis movement. We can, however, make some general observations. The shape of the geoid changes resulting from an axis shift is equal in magnitude and opposite in sign between the north and south hemispheres, with the maximum occurring at 45 degrees latitude. The shape of the geoid changes resulting from changes in the earth's rate of rotation, is symmetrical about the equator, with the maximum change occurring at the equator and pole (opposite signs). The shape of the geoid changes relating to gravitational attraction by the moon and sun, is nearly symmetrical about the equator The maximum occurs close to the moon and sun's nadir (near the equator). Gravitational tide amplitudes, however, are also modified by atmospheric pressure and wind.

Effect on the Magma Seas

All forces acting on the earth's oceans in the above discussion also act on the magma seas. The response, however, is considerably different due to the high viscosity of the magma seas, and they are encased in the earth's shell. If we try to pour thick catsup out of a bottle, it takes time to respond. The impatient customer that alternates between tilting a catsup bottle and setting it upright every 5 seconds will never get it to pour. For similar reasons, the crest of the magma tide, during the earth's rotation under the moon and sun, will only oscillate a small fraction of the 1-2 foot ocean tides. Likewise, the short term l.o.d. cycles do not provide sufficient time for the magma sea to respond completely. Even the 14.29 month cycle of the Chandler wobble may not offer enough time for the highly viscous magma seas to respond completely. Long term axis drifting, however, provides sufficient time for the magma seas to respond.

The magma tides (products of both gravitational pull and axis shift) are responsible for massaging the crust with a help from shifting atmospheric loading on the crust. High pressure magma at the crest of the magma tide intrudes into crustal fractures or bows up the crust. Sudden differential vertical movements between plates, under positive pressure, register as earth tremors. The low magma pressure during the pass of the magma tide trough allows the crust to settle. Sudden differential settling between adjacent plates is also registered as earth tremors. Normal crustal response, either up or down, introduces minute errors into tide gauges records.

Appendix F   THE AVOCATION SEED

The following is copied from the only written evidence of the theory developed by William Chester Strain. It was printed in pamphlet form, and addresses some of his thought process regarding the mechanics of an axis shift. As we will see, he considered the earth to be essentially rigid and attributed the effect of the changes in centrifugal forces as applying to the world oceans.

The Mechanics of the Movement of the Equator

(Copyright 1926, By W. C. Strain)

Upon the movement of the equator ninety degrees the North and South poles would sustain a flood of 13.3 miles, and the points on present equator that would become new North and South poles would sustain 13.3 miles drainage.

Should the equator move less distance there would be two areas of flood and two of drainage of less height and depth.

More than 500 years before Christ, Pythagoras, summing up the astronomical knowledge of the Chaldeans, Egyptians and Greeks, imagined a solar system with its central sun and revolving planets.

Claudius Ptolemy, over 600 years later, taught the popular theory of a flat Earth, the center of the universe and the sun passing round the earth in a night and a day, and we still say the sun rises in the east and sets in the west, though we know it does not.

And it was not until 1,400 years later and after Columbus discovered America, that Copernicus revived the theory of Pythagoras, stripped it of its embellishments and gave us the theory of the solar system as it is taught today.

Three years after the death of Copernicus, Tyco Brahe was born and Kepler, a student under Brahe, accepted the theory of Copernicus and with Galileo laid the foundations on which Sir Isaac Newton built.

The year Galileo died Newton was born and Newton, the student, Professor of Trinity College, Member of Parliment [Parliament], Master of the Mint, President of the Royal Society, was able with this prestige to impress on the skeptical people of his day wild theories and revolutionizing facts, which we recognize today as law, gravitation, the mutation of the axis, and the bulge of water under the equator, and there, where Newton dropped the work, we stand after 200 years, and do not know what will happen when that flood thirteen miles deep moves with the equator; what continents will be overwhelmed and what great areas of drainage will be formed.

If one could strike a circle with a radius of 4 miles plus the equatorial radius of the earth, and another circle 20 miles less, one would find that all the elevations and depressions of the earth could be contained between them.

The upper line might be called the height of elevation, for the summit of Chimborazo would almost touch it, and the lower line might be called sea bottom, for sea bottom at the North pole would almost touch it also.

Sea level at the equator would be about 4 miles below the height of elevation and sea level at the pole would be more than 17 miles below the height of elevation, or 13.3 miles nearer the center of the earth than sea level at the equator.

The summit of Mt. Everest, thirty degrees north would be short about 2 miles of reaching the height of elevation, and the high plateau of Tibet is about one mile nearer the center of the earth than the valley of the Amazon.

The line of "true sphere" which is the line representing sea level of our world if the world had no rotation would be about 8.5 miles above sea bottom.

If the world's rotation should decrease to a dozen for the year the flood of water 13.3 miles deep, held under the equator by centrifugal force, would flow into the Arctic and Antarctic oceans and the sea would be about 8 miles deep at the North Pole and 4 miles deep at the South Pole, and the bulge at the equator and flattening at the poles would disappear.

The drainage at the equator would be over 6 miles deep and the climate of those hot regions would be changed to that of the summit of Mt. Everest. Central Africa, the Sahara desert and the valley of the Amazon would be great fields of ice and snow with rarified air unsuited to sustain life in man.

About forty-five degrees north and south of the equator there should be no change of sea level but climate would be modified by the lengthened day.

This comes within the philosophy of those that substitute time for catastrophe and the change should be dated in million years hence.

But there are changes greater than these that can happen within a month and no man can set the date thereof.

Such changes have taken place in the past with records geological, legendary and written, and perhaps the change will occur again.

When we substitute time for catastrophe we view the great changes of the world without fear, and say, "Oh, well, we will be dead long before then", and while all these records show catastrophe, such changes do not rupture the earth's crust and the human race has survived them all.

And this catastrophe is the movement of the equator with its flood of 13.3 miles.

The law of mechanics that govern the movements of the equator tells us that at the least movement of the equator the water of the earth is divided into four areas: two of flood and two of drainage. The drainage that brought the ice age in North America caused the water to recede until the land stood miles above sea level.

The great boulders of Iowa are a part of the records, as are the drumlin of New York and Massachusetts, and all the debris in the states between and if we read the scratches on the boulders right, they say the boulders were moved south by a great ice field and that the ice stopped here and melted and the water gouged out the bed of the Missouri and Ohio rivers and drained south into the Mississippi.

And as this gives the southern limit of ice it also gives the position of the North pole at that time at a place near the magnetic pole or twenty degrees south, and also the position of the old equator well down in South America, 1,200 miles south of its present position.

This gives St. Louis a latitude of sixty degrees north and the law of mechanics that govern the shift of the equator tell us that there was drainage of the sea around North America until it had an added elevation of 14,000 feet and there was a flood of 14,000 feet in the south half of the Western Hemisphere; a like flood in the north half of the Eastern Hemisphere and a drainage of 14,000 feet in the south half of the Eastern Hemisphere, draining a continent that almost connects Australia and South Africa and accounts for the migration of the Black race.

The 14,000 feet of flood and drainage given above is only true on the meridians one hundred degrees west and eighty degrees east and this measurement decreases east and west for eighty degrees to meridians of stability that pass half-way between old and new poles, and cross the equator at a point where old and new equators cross each other, also a great circle of stability passing half-way between the old and new equators and crossing the meridians of stability at the point where they all cross the equator. These two points or nodes, do not change sea level. The lines of stability may change, but resume their old levels. For such is the law of the shift of the equator.

The astronomical view of the shift of the equator is this:

· The plane of the earth's orbit remains constant to the sun.
· The angle of the earth's axis remains constant to the plane of the orbit.
· The earth atmosphere with its equatorial bulge and flattened poles remain constant to the axis.
· The earth water with its equatorial bulge and flattened poles remains constant to the axis.
· The solid earth mass changes its angle with the axis.
· Part of the earth's surface moving toward the equator is submerged, and the part that moves away is drained.
· Although it would be an astronomer's view, it conflicts in no way with the more common view that considers the equator and poles moving in unison to new places on the earth's surface.

The cause of the shift of the equator is unknown to us, but is probably caused by some force outside the earth. In a written record of a shift of the equator, that of Noah's flood, the movement required forty days to mature. The words, "the fountains of the great deep were broken up" are a true description of what occurred. The earth did not reach equilibrium, and the return movement required 110 days. Still, Noah remained in the ark a full year.

We should remember, however, that there is as much flood in the world today as there ever was, and that it is only the relocation of these floods that could cause disaster.

There is a small portion of the earth's surface under the equator that could not be flooded except by greater speed of rotation of the earth, but this land is subject to drainage that would cover it with snow and ice almost as destructive to human life as flood, but there would probably be plenty of time to descend to lower levels.

The lower levels would probably be true desert, recently drained ocean bed, without vegetation, but descending streams of fresh water would carry some seed from higher ground and in time the desert would be clothed in verdure. The food supply at first would be fish and birds.

To visualize the surface of the earth one might draw two circles from the same center, one slightly larger than the other, and the space between would represent 13.3 miles. The equatorial flood would always touch the upper line and the drainage at the poles would touch the lower line. No matter where on the earth's surface the equator and poles rested, there would always be two floods, one on each side of the world, and the deepest point of which would be marked by the equator and two areas of drainage of equal depth marked by the pole.

W. C. STRAIN
Grand Junction, Colorado

Glossary

The following glossary reflects the intended meaning of terms employed in this book. These definitions may be slightly different or more restrictive than those in dictionaries or other glossaries. A few newly coined words are included to present the ideas.

Ancestral Rocky Mountains	A mountain chain believed to have existed near the site of the present Rocky Mountains.
Anticline	*see* crustal folds
Aphelion	The point in earth's orbit furthest from the sun. *see* perihelion.
Archimedes' principle	A body in a fluid is buoyed up by a force equal to the weight of the fluid displaced.
Asteroids	Small celestial bodies that can occasionally collide with our planet.
Asteroid Theory	Louis Alvarez' theory on mass extinction events.
Asthenosphere	The non-brittle sphere below the lithosphere.
Atmosphere	The sphere of gasses (air) that surrounds our planet.
Axis movements	
Axis drift	A shift in the earth axis that is presently moving the North Pole toward New England at the rate of one degree in about 1.25 million years.
Axis shift	Any shift of the earth axis between two time points.
Axis surge	A term introduced here to explain some geological phenomena —a relatively sudden shift of the axis as opposed to the ultra slow evolutionary drift..
Axis wobble	A wobble of the earth axis that causes the North Pole to trace a circular path over a period of about seven years.
Axis-of-axis shift	An imaginary axis perpendicular to the earth spin axis and the shift meridian plane.
Basalt*see* flood basalts.	
Basement rocks	The igneous and metamorphic rock platform on which sedimentary rocks are deposited.
Batholiths	A crustal blister pushed up by the magma seas. *see* also laccolith.
Biosphere	All living forms of plant and animal life on earth.
Block mountains	Mountains formed by a block of earth's crust breaking free on both sides allowing it to move upward as a unit.
Calderas	A depression formed by the collapse of a volcano vent or vents making the basin much larger than the original volcano vents.
Canadian Shield	Large region of Canada stripped down to the igneous and metamorphic basement rock by the continental ice sheet.
Carbon-14 dating	A method for age determination, using radioactive decay rate of carbon.
Cartesian grid	A common rectangular x-y grid.
Cataclysmal evolution	A concept that earth's evolutionary changes are triggered by cataclysmic events as opposed to uniformitarian evolution.
Cataclysmal Theory	A theory that proposes cataclysmic events mold earth's evolution.
Cells	A term used to represent units of circulating fluids in the atmosphere, hydrosphere and asthenosphere.
Centrifugal forces	The force directed away from the axis of any rotating body. The term "centrifugal force" as employed in this book primarily refers to the forces

	applied to the ocean waters and the magma seas caused by earth's rotation.
Challenger Deep	The deepest point in the oceans—35,958 feet below sea level in the Mariana Trench southwest of Guam.
Chandler Wobble	A cyclic movement of the earth's spin axis similar to the wobble of a spinning top. *see* axis movements.
Classic negative quadrant features	Geologic features formed by the crust subjected to compressive forces.
Classic positive quadrant features	Geologic features formed by the crust subjected to tensile forces.
Coal age	Since coal is normally assumed to have been the product of dense tropical vegetation, the term coal age in this book usually represents the opposite of ice age: a warm wet climate.
Coast and Geodetic Survey (C&GS)	A government agency responsible for establishing the geographic reference system used in surveying our planet.
Colorado Front Range	The eastern edge of the Rocky Mountains.
Computer modeling	Using high speed computer programs to analyze masses of data.
Conduction	Heat transferred through a solid body; heat escaping from earth's interior.
Continental drift	The concept advanced by Alfred Wegener that large continental plates move over the earth's interior. This concept is now incorporated into the Plate Tectonics Theory.
Convection	Heat transferred by gasses or fluids; atmosphere and ocean currents.
Coriolis effect	The horizontal circulation patterns produced in the atmosphere and hydrosphere by earth's rotation; the natural forces that spin whirlpools counterclockwise in the Northern Hemisphere and clockwise in the Southern Hemisphere.
Continental glaciers	Ice sheets or ice caps that cover a substantial area of a continent.
Conveyor Belt Theory	A theory proposing that tectonic plates are carried on a vertical cell of magma currents that upwell along midocean ridges, and downwell on the opposite side of the tectonic plate.
Crustal massaging	*see* massaging of the earth's crust.
Crevices	Cracks opened in the crust as it is bowed over a magma sea blister.
Crustal forces	Forces responsible for altering the earth's crust.
Compression	Refers to a notion that lower magma seas in the negative quadrants places the earth's crust under compression as it must fit within a shorter span; crustal forces that facilitate crustal folding, overthrust belts, subduction zones, etc.
Tension	Refers to a notion that higher magma seas in the positive quadrants places the earth's crust under tension as it must stretch to cover a longer span; forces that facilitate the formation of dikes, sills, batholiths, midocean ridges, etc.
Shear	Any differential within or between areas of crustal compression and crustal tension; forces responsible for all earthquakes and transverse fracture zones.
Crustal elevator	A notion that raising and lowering of the magma seas explains crustal movements between periods of sedimentary deposition and periods of erosion, for regions like the Colorado Plateau.
Crustal folds	Sedimentary rock layers distorted by tremendous forces.
Anticline	Convex fold
Monocline	Tilted

Syncline	Concave fold
Crustal fractures	*see* faults
Crustal gymnastics	Refers to forces that remodel earth's crust..
Crustal subsidence	*see* negative quadrants
Crustal rocks	*see* lithosphere.
Crustal uplift	*see* positive quadrants.
Deluge	Biblical flooding of the land.
Dikes	A geologic feature formed when molten magma pushes up into crustal fractures and freezes; vertical or near vertical intrusions are called dikes, and the horizontal or near horizontal intrusions are called sills.
Dinosaur age	The geologic period when dinosaurs inhabited earth; ending with a mass extinction event about 65 million years ago at the end of the Cretaceous Period.
Discontinuities	Evidence that periods of erosion have destroyed part or all of a geologic record of sedimentary deposits—leaving gaps in the geologic calendar.
Ductile	Refers to rocks below the brittle crust that can be deformed without fracturing.
Dynamic axis theory	The theory introduced in this book that suggests movements in earth's spin axis play a vital part in earth's evolution.
Dynamic balance	A spinning object (earth) with weight distributions such that no axis wobble exists.
Earth's axis-	
geodetic	Earth's spin axis as defined by a best-fit spheroid used as a geodetic reference system.
instantaneous	Since we now know that earth's spin axis wobbles and drifts, the instantaneous earth's axis refers to its position at a specific time point. An axis shift is the tilting of the axis between two instantaneous axis positions.
Earth's diameter	
equatorial	Diameter measured across an equatorial plane; approximately 27 miles greater than the polar diameter.
polar	Diameter measured from pole to pole.
poles	The intercept of the earth's axis with the earth's outer surface.
radius	The radius varies as a function of latitude; decreasing as the latitude increases.
Earth's dynamic balance	A theoretical condition that assumes the weight distribution is such, that it spins smoothly on its axis, without axis wobble or drift.
Earthquakes	Any rupture of the earth's crust; identified by measuring the resulting seismic vibrations (waves).
Electromagnato- meter	An instrument used to measure the electrical emissions from within our planet.
El Niño	A shifting warm water current in the Pacific credited with altering weather patterns. La Niña is its cold water companion.
Elevation	The distance above or below sea level. see geoid.
Equatorial bulge	Refers to the increase in earth's radius created by centrifugal forces thrusting the ocean waters toward the equator.
Equipotential surface	Refers to the balance of forces that establish sea level. *see* geoid.
Evolution	Change over geologic time.
Faults	Fractures in the earth's crust where slippage occurs during earthquakes.
New Madrid	A major north-south fault along the Mississippi Valley. The strongest

earthquakes ever recorded in the United States occurred near New Madrid, Missouri in 1812.

San Andreas
A major north-south fault along the west coast created by the slippage of the Pacific Plate relative to the North American Plate.

Transverse Fault Zones
A family of parallel crustal fractures perpendicular to midocean ridges.

Fault types

Dip or dip slip
A fault with its relative slippage perpendicular to the major dimension of the fault surface; normally occurs on oblique faults.

Strike or strike slip
A fault with its relative slippage parallel to the major dimension of the fault surface; normally occurs on near vertical faults.

Fjords
Valleys carved by glaciers that are now below sea level; some over 4,000 feet below sea level.

Flat earth theory
A popular theory before astronomy and geodesy proved the earth is orb-shaped.

Flood basalts
Magma (lava) that flows over earth's surface and freezes into basalt rock.

Force couple
Two forces separated by some distance (moment arm) acting in opposite directions on a body.

Foreign terranes
Blocks of tectonic plates torn free from their parent plate and welded to another plate.

Fossils
Evidence of ancient life preserved when minerals have replaced the original body.

Fracture zones
Areas of multiple faults.

Friction lock
The resistance to slippage between two blocks; specifically the opposing surfaces of a crustal fault.

Frozen mammoths
Mammoths quickly frozen and preserved for thousands of years.

Geocentric
A X, Y, Z, geodetic reference system with its origin at the earth's center.

Geodetic reference systems
Mathematically derived reference planes used to define the size and shape of our planet; the foundation of all surveys.

Geoid
A geodetic term for sea level; an equipotential surface that undulates as a function of variations in gravity.

Geodesist
A professional involved in the measurement of the size and shape of our planet.

Geologist
A professional involved in the physical construction of our planet.

GEOSAT
A navy satellite launched by the NASA that measured its distance above the planet; recorded undulations in sea levels using downward looking radar. *see* also SEASAT.

Glacial erratic
Large rocks foreign to an area that were transported and deposited by glaciers.

Gondwana
A supercontinent composed of Southern Hemisphere continents.

Greenwich meridian
The meridian of zero degree latitude passing through Greenwich, England.

Gravimetric surveys
Surveys measuring variations in gravity at the surface of our planet.

Guyots
A flat toped seamount. *see* Seamounts.

Hot spots
Volcanic regions where the magma has pushed up into the crust. Often refers to hot spots such as the one presently under Hawaii that has left a trail of volcanic scars (seamounts) in its path.

Hypothesis
A theory that has been elevated to an assumed truth for the purpose of explaining another theory.

Hydrocarbons	A primary building block of all plant and animal forms.
Hydrologic cycle	The cycle of ocean water evaporation, clouds, precipitation, and its return to the oceans through the rivers.
Hydrosphere	The oceans, glaciers, and other waters of the world.
Ice age	A synonym for a period of extensive glaciation.
Ice cap	A large glacier that is thick enough to completely hide any rock outcrops or peaks.
Ice sheet	A glacier that covers a large segment of a continent.
Igneous rocks	Rocks formed when earth's magma freezes: from very fine grained glassy obsidian to crystalline granites.
Insolation	Refers to the amount of sun reaching our planet.
Isostasy	Natural laws governing floating bodies: the weight of the fluid displaced equals the weight of the floating body.
Isostatic balance	Refers to the crust's natural response as a body floating on the magma seas. Example: North America sank deeper into the magma seas when it was covered by the continental ice sheet.
Isothermal- pressure balance	Refers to the heat and pressure condition that define the MOHO.
Kinetic energy	The energy possessed by the momentum of earth's rotation.
Laccoliths	A crustal blister caused by magma pushing up through an orifice before spreading out between layers of sedimentary rocks; similar to a batholith.
La Niña	*see* El Niño.
Lateral moraines	Ridges of rocks and debris eroded by a glacier and deposited along the outer edge of its path.
Latitude	The angular distance of a point on our planet north or south of the equator.
Laurasia	A supercontinent formed by continents of in the Northern Hemisphere; the companion of Gondwana in the Southern Hemisphere.
Laws of dynamics	Predictable reactions of bodies being acted upon by outside forces.
Literary license	Allowance for technical exactness in fiction writing.
Lithosphere	The thin (rock like) outer layer of our planet
Longitude	The angular distance of a point on our planet east or west of the Greenwich Meridian.
Magma	A general term referring to the molten interior of our planet. When cooled it becomes igneous rocks.
seas	Refers to the earth's interior below the lithosphere that can be remolded without fracturing.
sea currents	The movement of magma within our planet as they respond to tempera ture differences and the centrifugal forces of earth's rotation.
void	Refers to the condition that temporarily exists when magma flows away from negative quadrants and allows the crust to subside.
tide	Tide movements in the magma seas caused by changes in centrifugal forces acting on the magma seas and the cyclic variations of moon and sun's gravitational pull.
tornadoes	A term suggested as an explanation for hot spots as the magma sea's equivalent of atmospheric tornadoes.
Magnetic poles	Points of maximum magnetic strength. The north magnetic pole is presently positive while the south magnetic pole is negative. Both

magnetic poles are a considerable distance from the spin axis North and South Poles. Both drift over time and occasionally reverse their polarity: south becomes positive and north becomes negative.. Compasses point to the magnetic North Pole.

Magnetic pole reversal	*see* magnetic poles
Magnetic signatures	The orientation of magnetic forces frozen into rocks when they were formed; indicating declination and dip (direction) toward the north magnetic pole.
declination	Horizontal angle deviation from North.
dip	Vertical angle deviation from level.
paleomagnetic data	Records derived from rocks.
isomagnetic charts	Charts using contours to show lines of equal magnetic signatures.
Magnetosphere	The earth's magnetic field.
Mantle-core boundary	The surface between earth's lower mantle and its denser core.
Marine terraces	Benches, some hundreds of feet above or below sea level that display scars of beach erosion from ancestral times.
Massaging the earth's crust	Forces applied to the earth's crust as a result of cyclic shifting of the magma seas; magma seas responding to axis movements and gravitational pull of the moon and sun.
Mass extinctions	Periods in our earth's history when complete species died out. Usually attributed to extreme climate swings.
Maunder Minimum	A low point in a cyclic variation in sun's intensity suggested by E. Walter Maunder as being the cause of the "Little Ice Age" in the 16th to 18th centuries.
Mercator projection	A map projection often used for world maps that distorts latitude lines to the North and South of the equator. The east-west dimension of arctic lands are greatly stretched, distorting their shape.
Meridian	The north-south lines of longitude.
Metamorphic rocks	Igneous or sedimentary rocks that have been subjected to tremendous heat and pressure; creating a new type of rock.
Meteorologist	A scientist involved in weather forecasting.
Midocean ridges	Ridges that pass through all of earth oceans where new crust is forming as the tectonic plates pull apart.
Milankovitch Cycles	A theory attempting to explain earth's climate swings as a function of the long term changes in the orientation of earth's axis relative to the sun. Suggested as a key factor in ice age cycles.
Modified Mercalli Scale	A scale devised to indicate the magnitude of damage caused by earthquakes. *see also* Richter Scale.
MOHO	Mohorovicic discontinuity, the boundary between the brittle crust and the mantle.
Moment (mechanics)	An eccentric force acting on a body; value determined by multiplying the force by the moment arm (leverage).
Moment arm	The distance between the opposite directed forces of a force couple; creating leverage.
Monocline	*see* crustal folds.
Monumented	Fixed reference points established by geodetic surveys to

bench marks	determine latitude, longitude and elevation; or simply elevation.
Nadir	Point directly below; opposite of zenith.
Neutral equatorial great circle	A term coined for this book to define a line that theoretically will have no change in sea level due to an axis shift; a line midway between the pre shift and post shift; equators.
Neutral meridional great circle	A term coined for this book to define a line perpendicular to the neutral equatorial great circle that theoretically will have no change in sea level due to the shift; meridians at 90 degrees from the shift meridian.
North American datum horizontal	Clark's 1866 spheroid
vertical 1929	Datum used to establish elevations.
vertical 1988	Datum reflecting latest geodetic adjustment.
Nuclear ovens	Heat producing nuclear activity deep within earth's interior; considered a prime driver of magma currents.
Oblate ellipsoid	A mathematically perfect ellipsoid created by rotating an ellipse about its axis.
prolate ellipsoid	Spin axis diameter greatest (football).
oblate ellipsoid	Equatorial diameter greater than polar diameter (earth).
Ocean currents	Natural current patterns within earth's oceans.
tides	Fluctuation in sea levels primarily due to gravitational pull by the moon.
trenches	Deep gashes in the earth's surface where one tectonic plate is being subducted back into the earth's interior.
Orbit-earth's	An elliptical path around the sun.
Orbit-moon's	An elliptical path around the earth.
Orifice vent	The crustal opening that allows magma to move up through the crust.
Oscillations	Wave type motion.
Overthrust belt	A geologic feature created by one tectonic plate sliding over an adjacent tectonic plate. The overriding plate remains sandwiched with the lower plate, as opposed to a subducting plate that dives at a steep angle into the earth.
Oxygen isotopes	Isotopes used by scientists to estimate ancestral temperatures.
Pangea	A supercontinent made up of all continents which is believed to have existed 200 to 300 million years ago.
Perihelion	The point in earth's orbit closest to the sun; occurs in January. *see* aphelion.
Plate edges	Referring to the interplay between tectonic plates.
Precession	A 25,000 year cyclic change in the orientation of earth's axis relative to the stars that alters the seasonal distribution of sun's heat.
Quadrants	
positive	A term introduced in this book; the two quadrants with decreased latitudes causing the ocean waters to shift and produce higher sea levels (continental flooding). The same quadrants have higher magma seas that places the crust under tension.
negative	The opposite quadrants with increased latitudes causing the ocean waters to shift and produce lower sea levels (exposing some ocean floors). The same quadrants have lower magma seas that places the crust under compression.
Radiation	Heat transferred by atomic particles or rays; heat from the sun.
Radioactive elements	Elements that decay at known rates. The ratio of the decayed element to

	its original state is used to estimate geologi-cal time scales.
Richter Scale	A logarithmic scale used to measure the intensity of earthquakes. *see* Modified Mercalli Scale
Rigid crust	A term employed in this book (only for the purpose of explaining how ocean waters respond to an axis shift) that considers the lithosphere to be rigid and unresponsive to an axis shift.
Ring of Fire	Refers to the line of active volcanic activity along the Pacific Plate edges.
Sawtooth mountains	Mountains formed when a series of crustal blocks are tilted upward.
Scientific community	A general term referring to all earth scientist.
Seamounts	Volcanic mounds on the ocean floor.
SEASAT	NASA's satellite with radar altimeter to measure changes in the open sea surfaces. *see* also GEOSAT.
Sedimentary beds	Layers of sediments deposits identifiable as coming from the same gener al source and under essentially constant conditions; used by geologist to identify geologic ages.
Sedimentary recycling	A concept presented in this book that recognizes that most sediments transported and deposited in new sedimentary bed come from existing sedimentary beds. Only a token contribution comes from the harder igneous and metamorphic rocks.
Sea level	Geodesists reference plane for determining elevations above or below sea level; mathematically derived from a network of tide gauge stations, and extended into continent interiors by differential leveling.
Seismic waves	Earthquake vibrations used by scientists to measure the earthquake's intensity and to locate its epicenter.
Shear forces	*see* Crustal forces.
Shift meridian	A meridian along which the poles travel for any actual of theoretical axis shift; two meridians 180 degrees apart..
Sills	*see* dikes.
Spheroid	A best-fit mathematically perfect reproduction of our planet. *see* oblate ellipsoid.
Spreading zones	Refers to the midocean ridges where plates are spreading and new crust is created.
Subduction zones	The deep ocean trenches where one tectonic plate is subducted back into the earth's interior, thus shortening the crust. A companion feature of midocean ridges that lengthens the crust.
Submarine canyons	Geologic feature equivalent to land eroded canyons that extend out from most major river deltas to depths thousands of feet below sea level.
Supercontinent	Refers to periods when continents were grouped. *see* Pangea, Gondwana, and Laurasia.
Syncline	*see* crustal folds
Tertiary period	The geologic period extending from about 65 million to 3 million years ago. A period characterized by the erosion of lands elevated after deposition of sediments during earlier periods.
Tectonic Plate Theory	A major theory introduced in the mid-twentieth century: addresses the division or our planet's crust into tectonic plates that move from each other. New crust is formed along midocean ridges where plates are pulling apart. Where plates collide the crust may either thrust up to create

a mountain range, or subduct back into the earth's interior. Plates may also slip past each other along their edge, such as the SanAndreas Fault.

Terminal moraines	Ridges of rocks and debris eroded by a glacier and deposited at its furthest point of advance.
Theory	A visualized but unproven concept.
Tides	
ocean	Shifting of the ocean waters in response to gravitational forces of the moon and sun.
earth	Shifting of the magma seas in response to gravitational forces of the moon, and sun, or changes in centrifugal forces caused by axis movements which in turn create vertical fluctuations of earth's crust.
gauge stations	Instruments that record the fluctuation in ocean tides for the purpose of establishing mean sea level.
Tomographic maps	Maps showing hot and cool areas of earth's crust.
Topographic maps	Maps that use contours to represent crustal relief.
Torques	*see* Force couple.
Transverse fracture zones	The near parallel crustal fractures associated with mid-ocean ridges.
Trenches	*see* Ocean trenches.
Tsunami	A tidal wave triggered by the vibration of an earthquake that can be very destructive on the opposite ocean coast.
Turbidity currents	Local fast moving currents within the oceans; the equivalent of high winds in the atmosphere.
Undulation	Refers to the variations in sea level from the mathematically best fit spheroid.
Uniformitarianism	A theory that assumes earth's evolutionary changes are gradual as opposed to cataclysmal evolution.
UTM (Universal Transverse Mercator)	A geodetic reference system used by the military.
Vector	The direction a force is applied.
component	The effective portion of a vector force; normal and tangential forces.
normal	The force acting perpendicular to the surface under consideration.
tangential	The force acting parallel to the surface under consideration.
sum	The direction a force is applied when considering the effect of more than one force vector.
Volcanic ash	Fine pyroclastic material fall out from a volcanic eruption.
Volcanic islands	Seamounts that extend above sea level.
Weight shifts	In this book, refer to any movement of a weight from one place to another on or within our planet.
Wobble cycle	The time for the axis wobble (North Pole) to complete one circular trace.
Zenith	Point directly above; opposite of nadir.

REFERENCES

Only a few of the following books or articles directly support the new the dynamic axis theory. The majority address today's vogue theories and are included with the suggestion that their ideas be evaluated from the perspective of the dynamic axis theory.

1. Anon., "Carbon-14 dating misses mark", *Rocky Mountain News*, 3 June 1990, p..29.
2. Anon., "Air pressure, earthquake link studied", *The Denver Post Wire Service*, The Denver Post, 15 May 1988, p. 6A.
3. Anon., "Asteroid hit linked with glacier age", *Denver Post*, 31 July 1988.
4. Anon., "Cuba proposed site for K/T impact", *Science News*, 1990, *137* (17), 268.
5. Anon., "Dino-death: flash broil or slow steam", *Science News*, 1990, 79.
6. Anon., "Meteorite may have triggered ancient, Texas-size wave", *Rocky Mountain News*, 29 July 1988, p. 46.
7. Anon., "Meteorologist ties air pressure to earthquakes", *The Daily Sentinel*, Grand Junction, 16 May 1988, p. 5A.
8. Anon., "Texas wave led to extinctions, CU prof says", *Rocky Mountain News*, 30 July 1988, p. 48.
9. Alvarez, W., "What Caused the Mass Extinction? *Scientific American*, 1990, *263* (4), 76. The above series served as an introduction to a debate between Alvarez and Courtillot, *see* Alvarez, W. and Asaro, F., "An extraterrestrial impact", and Courtillot, V., "A volcanic eruption".
10. Anderson, J.G.; Bodin, P.; Brune, J.N.; Prince, J.; Singh, S. K.; Quaas, R.; Onate, M.; "Strong ground motion from the Michoacan, Mexico earthquake", *Science*, 1986, *233*, 1043-1049.
11. Bakun, W.H. and Lindh, A.G, "The Parkfield, California, earthquake prediction experiment", *Science*, 1985, *229*, 619-624.
12. Bercovici, D.; Schubert, G.; and Glatzmaier, G. A., "Three-dimensional spherical models of convection in the earth's mantle", *Science*, 1989, *244*, 950-955.
13. Bohor, B. F.; Modreski, P. J.; Foord, E. E., "Shocked quartz in the Cretaceous-tertiary boundary clays: evidence for a global distribution", *Science*, 1987, *236*, 705-709.
14. Bonatti, E.; and Crane, K., "Oceanic fracture zones", *Scientific American*, 1984, *250* (5), 40.
15. Broecker, W. S., "The ocean", *Scientific American*, 1983, *249* (3), 146-160.
16. Broecker, W. S., and Denton, G. H., "What drives glacial cycles?", *Scientific American*, 1990, *262* (1), 48-56.
17. Brown, B. and Morgan, L., "The Miracle Planet", Gallery Books, New York, 1990.
18. Brown, H. A., "Cataclysm of the Earth", Freedeeds Associates, New York, 1968.
19. Bull, W. B., and Cooper, A. F., "Uplifted marine terraces along the Alpine fault, New Zealand", *Science*, 1986, *234*, 1225-1228.
20. Burchfiel, B. C., "The continental crust", *Scientific American*, 1983, *249* (3), 130-142.
21. Carter, W. E. and Robertson, D. S., "Studying the earth by vary-long-baseline interferometry", *Scientific American*, 1986, *255* (5), 46-54.
22. Cloud, P., "The biosphere", *Scientific American*, 1983, *249* (3), 176-189.
23. Cooke, R., "Caribbean likely site of asteroid hit", *The Denver Post*, 8 April 1990, p. 4A.
24. Courtillot, Vincent, and Besse, Jean, "Magnetic field reversals, polar wander, and core-mantle coupling", *Science*, 1987, *237*, 1140-1145.
25. Courtillot, V. E., "An extraterrestrial impact", *Scientific American*, 1990, *263* (4), 76-92.
26. Covey, C., "The earth's orbit and the ice ages", *Scientific American*, 1984, *250* (2), 58-66.
27. Dawson, J., ""CAT scanning the earth", *Earth*, May 1993.
28. Eckert, A. W., "The HAB Theory", Popular Library, New York, 1977.
29. Erickson, J., "Violent Storms", TAB Books Inc., Blue Ridge Summit, 1988.
    "Ice Ages: Past and Future", TAB Books, Inc., Blue Ridge Summit, 1990.

30. Foukal, P. V., "The variable sun", *Scientific American*, 1990, *262* (2), 34.

31. Francheteau, J., "The ocean crust, " *Scientific American,* 1983, *249* (3), 114-129.

32. Francis, P., "Giant volcanic calderas", *Scientific American,* 1983, *248* (6), 60-70.

33. Francis, P., and Self, S., "The eruption of Krakatau", *Scientific American*, 1983, *249* (5), 172-187.

34. Frezon, S. E.; Finn, T. M.; Lister, J. H., "Total Thickness of Sedimentary Rocks", Open file report 83-920, *United States Geological Survey,* Denver, 1963.

35. Goodman, J., "We Are The Earthquake Generation", Seaview Books, 1978, Berkley Books, New York, 1979.

36. Hallam, A., "End-Cretaceous mass extinction event: argument for terres-trial causation", *Science,* 1987, *238,* 1237.

37. Hapgood, C. H., "Earth's Shifting Crust", Pantheon Books, Inc., New York, 1958; "The Path of the Pole", Chilton, Philadelphia, 1978; "Maps of the Ancient Sea Kings, Evidence of Advanced Civiliization in the Ice Age", E.P. Dutton, New York, Revised American Edition, 1979.

38. Hoffman, K. A., "Ancient magnetic reversals: clues to the geodynamo", *Scientific American*, 1988, *258* (5), 76.

39. Horsfield, B., and Stone, P. B., "The Great Ocean Business", Coward, McCamm, Geophegan, New York, 1972.

40. Howell, D. G., "Terranes", *Scientific American*, 1985, *253* (5), 116-125.

41. Hsü, K. J., "When the Black Sea was drained", *Scientific American*, 1978, 238 (5), 52-63.

42. ___ "The Mediterranean was a desert: a voyage of the Glomar Challenger", Princeton University Press, Princeton, 1983.

43. *Ingersoll, A. P., "The atmosphere", *Scientific American*, 1983, *249* (3), 162-173.

44. Izett, G.A., Dalrymple, G.B., Snee L.W., "40Ar/39Ar age of Cretaceous-tertiary boundary Tektites from Haiti", *Science,* 1991, *252,* 1539-1541.

45. Jeanloz, R., "The earth's core", *Scientific American*, 1983, *249* (3), 56-65.

46. Jeanloz, R. and Lay, T., "The core-mantle boundary", *Scientific American,* 1993, *268* (5), 48-55.

47. Jablonski, D., "Background and mass extinctions: the alternation of macroevolutionary regimes", *Science,* 1986, *231,* 129.

48. Johnson, A. C. and Kanter, L. R. "Earthquakes in stable continental crust", *Scientific American,* 1990, *262* (3), 68-75.

49. Johnson, A. C., "A major earthquake zone on the Mississippi", *Scientific American,* 1982, *246* (4), 60-68.

50. Jones, D. L.; Cox, A.; Coney, P.; and Beck, M., "The growth of western North America", *Scientific American,* 1982, *247* (5), 70.

51. Jordan, T. H., and Minster, J. B., "Measuring Crustal Deformation in the American West", *Scientific American,* 1988, *259* (2), 48-58.

52. Kerr, R. A., "Were North Pacific waters sinking 18,000 years ago?", *Science,* 1985, *228,* 1519-1920.

53. ___"Continental drift nearing certain detection", *Science,* 1985, *229,* 953-955.

54. ___"A search for another San Andreas", *Science,* 1986, *231,* 116-117.

55. ___"Long Valley is quiet but still bulging", *Science,* 1986, *231,* 116.

56. ___"Parkfield earthquake looks to be on schedule", *Science,* 1986, *231,* 116.

57. ___ "Ancient river system across Africa proposed", *Science,* 1986, *233,* 940.

58. ___"Do California quakes portend a large one?" *Science,* 1986, *233,* 1039-1040.

59. ___"The big picture of the Pacific's undulations", *Science,* 1989, *243,* 739-740.

60. ___"Did the ocean once run backward?" *Science,* 1989, *243,* 243.

61. ___"Milankovitch climate cycles through the ages", *Science,* 1987, *235,* 973-974.

62. ___"Deep holes yielding geoscience surprises", *Science,* 1989, *245,* 468-470

63. ____ "Loma Prieta quake unsettles geophysicists", *Science*, 1989, *246*, 1562-1563.

64. ____ "Marking the ice ages in coral instead of mud", *Science*, 1990, *248*, 31.

65. ____ "Do plumes stir earth's entire mantle?', *Science*, 1991, *252*, 1068-1069.

66. ____ "The stately cycles of ancient climate", *Science*, 1991, *252*, 1254-1255.

67. ____ "Awaiting the next Mexico City earthquake", *Science*, 1987, *237*, 1118.

68. ____ "Did a volcano help kill off the dinosaurs?" *Science*, 1991, *252*, 1496.

69. ____ "European deep drilling leaves america behind", *Science*, 1989, *245*, 816-817.

70. Lambeck, K., "The Earth's Variable Rotation Geophysical Cause and Consequences", Cambridge University Press, Cambridge, 1980.

71. Lewin, R., "Mass extinctions select different victims", *Science*, 1986, *231*, 129.

72. ____ "Biologists disagree over bold signature of Nature", *Science*, 1989, *244*, *527*.

73. Levenson, T., "Ice Times: Climate, Science, and Life on Earth", Harper & Row, New York, 1989.

74. Levy, M., and Christie-Blick, N., "Pre-Mesozoic Palinspastic reconstruction of the eastern great basin, (Western United States)", *Science*, 1989, *245*, 1454-1461.

75. Macdonald, K. C. and Fox, P. J., "The mid-ocean ridge", Science in the 20th Century, Special Issue, 1991, 142.

76. McKenzie, D.P., and Rickter, F., "Convection currents in the earth's mantle", *Scientific American*, 1976, *235* (5), 72-89.

77. McKenzie, D.P., "The earth's mantle", *Scientific American*, 1983, *249* (3), 66.

78. Milne, Antony, "Earth's Changing Climate: The Cosmic Connection", Prism Press, Bridgeport, 1989, Avery Publishing Group Inc., New York.

79. Mogi, Kiyoo, "Recent earthquake prediction research in Japan", *Science*, 1986, *233*, 324-233.

80. Monastersky, R., "Inner space", *Science News*, 1989, *136*, 266-268.

81. ____ "Christmas quake presents geologic gift", *Science News*, 1991, *139*, 164.

82. ____ "Perils of prediction,: *Science News*, 1991, *139*, 376-379.

83. ____ "Earthquakes leaves bay area still vulnerable", *Science News*, 1989, 136 (17), 266-268.

84. Mossman, D. J., and Sarjeant, W. A. S., "The footprints of extinct animals", *Scientific American*, 1983, *248* (1), 74.

85. Munk and MacDonald, "Rotation of the Earth", Cambridge University Press, Cambridge, 1968.

86. Nance, R. D.; Worsley, T. R.; and Moody, J. B., "The supercontinent cycle", *Scientific American*, 1988, *259* (1), 72-79.

87. Powell, C. S., "Peering inward." *Scientific American*, 1991, *264* (6), 100-111.

88. Press, F., "Earthquake prediction, *Scientific American*, 1975, *232* (5), 14-23.

89. Raup, D. M., "Biological extinction in earth history", *Science*, 1986, *231*, 1528-1533.

90. Redfern, R., "The Making of a Continent", Times Books, New York, 1983.

91. Revelle, R., "The oceans", *Scientific American*, 1969, *221* (3), 54-65.

92. Ruddiman, W. F. and Kutzbach, J. E., "Plateau uplift and climate change", *Scientific American*, 1991, *264* (3), 66-75.

93. Russell, D. A., "The mass extinctions of the late Mesozoic", *Scientific American*, 1982, *246* (1), 58-65.

94. Sclater, J. G. and Tapscott, C., "The history of the Atlantic" *Scientific American*, 1979, *240* (6), 156-174.

95. Segall, P., and Harris, R., "Slip deficit on the San Andreas Fault at Parkfield, California, as revealed by inversion of geodetic data", *Science*, 1986, *233*, 1409-1413.

96. Siever, R., "The dynamic earth", Scientific American, 1983, 249 (3), 46-55.

97. Stanley, St. M, "Extinctions - or, which way did they go", *Earth*, 1991, *1* (1), 18-27.

98. Stewart, R. W., "The atmosphere and the ocean", *Scientific American*, 1969, *221* (3), 76.

99. Stommel, H., and Stommel, E., "The year without a summer", *Scientific American*, 1979, *240* (6), 176-186.

100. Velarde, M. G., and Normand, C., "Convection", *Scientific American*, 1980, *243* (1), 92-108.

101. Velikovsky, I., "Earth in Upheaval", Doubleday, New York, 1955.

102. ___"Worlds in Collision", Pocket Books, New York, 1977.

103. Vink, G. E.; Morgan, W. J.; and Vogt, P. R., "The earth's hot spots", *Scientific American,* 1985, *252* (4), 50-57.

104. Ward, C. M., "New Zealand Marine Terraces: Uplift Rates", *Science,* 1988, *240,* 803.

105. Weiner, J., "Planet Earth", Bantam Books, New York, 1986.

106. Wesson, R. L.; and Wallace, R. E., "Predicting the next great earthquake in California", *Scientific American*, 1985, *252* (2), 35-43.

107. Wetherill, G. W., "The formation of the earth from planetesimals", *Scientific American,* 1981, *244* (6), 162-174.

108. White, J., "Pole Shift", Doubleday 1980, Berkley Books, New York, 1982.

109. White, R. S., and McKenzie, D. P., "Volcanism at rifts", *Scientific American*, 1989, *261* (1), 62-71.

110. Wilson, J. T., "Continental drift", *Science in the 20th Century,* Special Issue, 1991, 114.

111. Winograd, I. J.; Szabo, B. J.; Coplen, T. B.; and Riggs, A. C., "A 250,000-year climatic record from great basin vein calcite: implications for Milankovitch theory", *Science,* 1988, *242,* 1275-1279.

Subject Index

To Order

additional copies of this book, other bestsellers in this *Frontiers in Astronomy and Earth Science* series, or other ATL Press professional science titles, please contact:

ATL PRESS, Inc., Science Publishers
Distribution Center, P.O. Box 4563 T Station, Shrewsbury, MA 01545, USA
ORDER TOLL FREE AND USE YOUR CREDIT CARD (all major cards)
Tel: 1-800-835-7543 (USA & Canada); 1-(508)-898-2290
FAX: 1-(508)-898-2063; E-Mail 104362,2523@compuserve.com

THE COMET HALE-BOPP BOOK
Guide to an Awe-Inspiring Visitor from Deep Space
Thomas Hockey
Foreword by comet co-discoverer Thomas Bopp
Frontiers in Astronomy and Earth Science, Volume 1

This is the comet guide everyone has been waiting for. The gripping narrative helps you prepare for what could well be this century's celestial event. This fascinating book provides essential star charts to locate the comet through July 1997. Hockey, a NASA Shoemaker-Levy 9 investigator, describes the comet's discovery and astronomer's preparation for its arrival. He guides you on how, where and when to observe the comet. This celebrated book, already featured by three science book clubs, is lavishly illustrated with photos and diagrams. It explains in plain terms the science behind comets. Hockey's lucid and compelling style stimulates comet buffs and converts young and mature novices into backyard observers. Internationally-recognized, Dr. Hockey is a professional astronomer, distinguished author, and experienced science educator.

1996, 176 pages, illustrated, 6 x 9
ISBN 1-882360-14-1, 1996, Cloth US $42.90 (Outside US $52.50)
ISBN 1-882360-15-X, 1996, Paper US $19.95 (Outside US $25.50)

COMET HALE-BOPP T-SHIRT
COMET HALE-BOPP-THE COMET OF THE CENTURY T-SHIRT
THE YEAR OF THE COMET, 1996-1997 T-SHIRT

each $15.95 (100% cotton, assorted colors, sizes S, M, L, XL)

THE EARTH'S SHIFTING AXIS
Clues to Nature's Most Perplexing Mysteries
Mac B. Strain
Frontiers in Astronomy and Earth Science, Volume 2

1997, 288 pages, illustrated, indexed, 6 x 9
ISBN 1-882360-30-3, Cloth US $42.95 (Outside US $52.50)
ISBN 1-882360-31-1, Paper US $19.95 (Outside US $29.50)

THE CASE FOR SPACE
Who Benefits from Explorations of the Last Frontier?
Paul S. Hardersen
Frontiers in Astronomy and Earth Science, Volume 3

"...a rare and important book...it shows how space technology makes very personal contributions to the quality of life of people."
    Scott Pace, Executive VP, National Space Society, Chairman, Policy Committee

"This book is long overdue...Read this book and then mail it to your congressman. Hardersen's is a message they need to hear on Capital Hill."
    Dr. Robert Zubrin, Chairman, Executive Committee, National Space Society

"..a compelling, easy-to-read argument for space exploration...appealing to both the novice and the expert...when you finish the book, you have a clearer picture of the potential of space technology."
    US Senator Chuck Grassley

1997,194 pages, illustrated, indexed, 6 x 9
ISBN 1-882360-47-8, Cloth US $42.95 (Outside US $52.50)
ISBN 1-882360-48-6, Paper US $19.95 (Outside US $29.50)

Frontiers in Biomedicine and Biotechnology
VOLUME 1 CARBOHYDRATES AND CARBOHYDRATE POLYMERS
        M. Yalpani, Editor   ISBN 1-882360-40-0, 1993
VOLUME 2 LEVOGLUCOSENONE & LEVOGLUCOSANS
        Z. J. Witczak, Editor   ISBN 1-882360-13-3, 1994
VOLUME 3 BIOMEDICAL FUNCTIONS AND BIOTECHNOLOGY OF
        NATURAL AND ARTIFICIAL POLYMERS
        M. Yalpani, Editor   ISBN 1-882360-02-8, 1996

Frontiers in Foods and Food Ingredients
VOLUME 1 SCIENCE FOR THE FOOD INDUSTRY OF THE 21ST CENTURY
        M. Yalpani, Editor   ISBN 1-882360-45-1, 1993
VOLUME 2 NEW TECHNOLOGIES FOR HEALTHY FOODS &
        NUTRACEUTICALS
        M. Yalpani, Editor   ISBN 1-882360-10-9, 1997